Density Functional Theory

Roshan Thotagamuge
Editor

Density Functional Theory

Foundations and Applications in Material Science

Springer

Editor
Roshan Thotagamuge
Department of Optometry
Faculty of Allied Health Sciences
University of Sri Jayewardenepura
Nugegoda, Sri Lanka

ISBN 978-981-95-9744-4 ISBN 978-981-95-9745-1 (eBook)
https://doi.org/10.1007/978-981-95-9745-1

This Springer imprint is published by the registered company Springer Nature Singapore Pte Ltd.
The registered company address is: 152 Beach Road, #21-01/04 Gateway East, Singapore 189721, Singapore

This book is dedicated to students, educators, and early-career researchers who seek to understand the fundamental principles governing the behavior of matter at the atomic and electronic scales. It is especially dedicated to undergraduate students of physics, chemistry, materials science, and related disciplines, whose curiosity and enthusiasm continue to drive the advancement of computational approaches in modern science. I also dedicate this work to all teachers and mentors who inspire learners to bridge theory and computation in the pursuit of scientific discovery.

Preface I

Density functional theory (DFT) has become one of the most powerful and widely used theoretical frameworks for understanding and predicting the properties of materials at the atomic level. Over the past few decades, DFT has transformed research across physics, chemistry, and materials science, enabling insights into electronic structure, bonding, defects, surfaces, and nanoscale phenomena that were once inaccessible.

Despite its broad impact, students and newcomers often find DFT challenging due to its strong foundations in quantum mechanics and its computational complexity. *Density Functional Theory: Foundations and Applications in Material Science* addresses this gap by presenting DFT in a clear, structured, and pedagogically sound manner. The book carefully guides readers from the fundamental concepts of quantum mechanics to modern DFT implementations and advanced applications.

This book is particularly well suited and highly recommended as a **textbook for undergraduate courses** such as *Computational Physics*, *Computational Materials Science*, and *Computational Chemistry*, while also serving as a **practical guide for research projects at early stages**. It is especially valuable for students and researchers from diverse disciplinary backgrounds who seek both fundamental and advanced knowledge of DFT. By combining theoretical clarity with practical examples and interpretation strategies, the book equips readers with the confidence to apply DFT tools meaningfully in both coursework and research.

I am confident that this book will serve as a valuable resource for students, instructors, and researchers seeking a balanced introduction to both the principles and applications of DFT.

Nugegoda, Sri Lanka

Roshan Thotagamuge

Preface II

Density functional theory (DFT) occupies a central position in modern computational science, offering a practical and accurate framework for studying the electronic structure of atoms, molecules, and solids. As computational power and software accessibility have increased, DFT has become an essential tool not only for researchers but also for undergraduate students engaging in computational modules and research projects.

This book was written with the primary aim of providing a **clear, coherent, and accessible introduction to DFT**, bridging the gap between abstract quantum mechanics and real-world materials applications. The content is organized to support **undergraduate students in physics, chemistry, materials science, and related disciplines**, particularly those enrolled in modules such as *Computational Physics*, *Computational Materials Science*, and *Computational Chemistry*.

The book is divided into thematic parts that progressively develop the reader's understanding. Early chapters introduce the motivation for DFT and its theoretical foundations, including the Hohenberg–Kohn theorems and Kohn–Sham formalism. Subsequent chapters focus on computational methods, commonly used software tools, and best practices for performing reliable DFT calculations. Later chapters explore applications in solid-state physics, surfaces, nanomaterials, defects, and emerging beyond-DFT approaches.

Special emphasis is placed on interpreting DFT results, an aspect often overlooked by beginners. In addition, this book is designed to function as a guidebook for undergraduate research projects across disciplines beyond chemistry, physics, and materials science, assisting students in designing simulations, analyze outputs, and connect computational results with physical and chemical insights.

I hope that this book will empower research and students to confidently engage with DFT, foster independent computational thinking, and serve as a stepping stone toward advanced research in materials science and related fields.

Nugegoda, Sri Lanka

Roshan Thotagamuge, Ph.D.

Acknowledgments

As the editor of this book, I would like to express my sincere gratitude to all chapter authors for their valuable contributions, dedication, and collaborative spirit in bringing this book to completion. Their expertise and commitment have been instrumental in shaping this volume into a comprehensive and coherent resource.

I am grateful to Springer for their professional guidance and support throughout the publication process. I also acknowledge the academic institutions, colleagues, and research communities that have fostered an environment conducive to learning, teaching, and research in computational science.

Special thanks I extended to students whose questions, curiosity, and research efforts inspired the pedagogical focus of this book. Their engagement in undergraduate coursework and research projects highlighted the need for a structured and accessible guide to density functional theory.

Finally, I acknowledge the support of my family and mentors, whose encouragement and understanding made this work possible.

Contents

Chapter 1
From Quantum Mechanics to Density Functional Theory

Visal de Silva, Azeez Ahamed, Piumantha Samaranayake, Nadeesha Manohari Wickramage, and Roshan Thotagamuge

Abstract This chapter described the theoretical developments of first-principles quantum mechanics parallel to many-electron systems focused on DFT. It starts from the Schrödinger equation and describes its usefulness in single-electron systems, touches on the insurmountable computational difficulties that emerge when electron–electron interactions are included for multi-electrons which is due to curse of dimensionality. In order to address these difficulties, various approximation techniques are developed such as Hartree or Hartree–Fock methods which approximate the many body problem by mean field and post-Hartree–Fock methods including Configuration Interaction or Coupled Cluster approaches where electronic correlation can be included more accurately. Although wavefunction-based approaches are very powerful, their computational scaling is not favorable and it hinders their application on large and complex systems. This shortcoming motivates an ongoing tendency to density-based methods, where the electron density is instead the main property. The chapter, for this reason, develops the concept of electron density and expresses the Hohenberg–Kohn theorems that are based on which DFT rests. We further describe the Kohn–Sham formalism as a means to practical realization of DFT for efficient and accurate electronic structure calculations. Overall, the chapter provides a coherent conceptual bridge between traditional quantum mechanical methods and modern density functional approaches widely used in computational chemistry, condensed matter physics, and materials science.

Keywords Density Functional Theory (DFT) · Many-Electron Systems · First-Principles Quantum Mechanics · Hartree–Fock and post-hartree–fock methods · Electron density · Kohn–sham formalism

---

V. de Silva (✉) · A. Ahamed · P. Samaranayake · N. M. Wickramage
Department of Physics, Faculty of Science, University of Ruhuna, Matara, Sri Lanka
e-mail: desilvavisal@gmail.com

R. Thotagamuge (✉)
Department of Optometry, Faculty of Allied Health Sciences, University of Sri Jayewardenepura, Nugegoda, Sri Lanka
e-mail: roshan@sjp.ac.lk

R. Thotagamuge (ed.), *Density Functional Theory*,
https://doi.org/10.1007/978-981-95-9745-1_1

1.1 Introduction

Quantum Mechanics, a masterpiece of human ingenuity has revolutionized our understanding of the electronic behavior in atoms and molecules. The introduction of probabilistic nature of electron positioning, governed by Schrödinger equation serve as an evidence for this revolution. It provides an accurate description for single-electron systems such as hydrogen atom. However, the Schrödinger equation faces significant limitations when it was applied to systems with many electrons such as large molecular structures. Since, large molecular structures possess numerous electron-electron and electron-nucleus interactions, this leads to an extremely complicated and intractable computational process. To address this issue, the necessity of developing approximation methods was put forward. One such methods is density functional theory (DFT). DFT is a powerful and an efficient approach, which provides solution to the many-body problem. Unlike traditional wavefunction-based methods, which attempt to solve for the full many-electron wavefunction, DFT reformulates the problem in terms of the electron density, a function of only three spatial coordinates rather than the exponentially large wavefunction space. This shift significantly reduces computational costs while still capturing essential quantum mechanical effects [1].

DFT has since become an indispensable tool in modern computational chemistry and condensed matter physics, enabling the study of electronic structures, molecular interactions, reaction mechanisms, and material properties with remarkable efficiency. In this chapter, we will explore the limitations of traditional quantum mechanics in many-electron systems, the development of approximation methods, and the fundamental concept of electronic density that underpins DFT.

1.2 Limitations of Traditional Quantum Mechanics for Many-Electron Systems

1.2.1 The Theoretical Framework and Many-Body Problem

In the early twentieth century, where the roots of quantum mechanics trace back to physicists like Max Planck, Albert Einstein, and Neils Bohr, they were building a new framework to explain phenomena that classical mechanics failed to address. Planck worked on blackbody radiation, which led to the discovery of quantization of energy, while Einstein extended this idea to explain the photoelectric effect, introducing the concept of photons. These discoveries laid the foundation for the building and the revolutionizing quantum mechanics.

The quantum revolution was fully realized with Schrödinger's wave mechanics in 1926, where he formulated the famous Schrödinger equation, which describes the wavefunction of a system [2]. Even though, the equation was accurate in describing a single particle system, it becomes incredibly complex in describing many-particle systems. It was soon recognized that the interactions between particles, especially electrons, in a multi-body system were not easy to describe, setting the stage for the challenges we face when dealing with many-body quantum systems.

1.2.2 *The Curse of Dimensionality*

In a system, as the number of particles increases, the number of dimensions required to describe it also increases. For a system with N number of electrons, the many-electron wavefunction is a function of the positions and spins of all N number of electrons. This dependence causes an explosion in the number of dimensions as N increases, making it practically impossible to solve the Schrödinger equation directly for large systems.

The many-electron Hamiltonian that describes the total energy of an interacting electron system is given by

$$\hat{H} = \sum_i \left[\frac{\hbar^2}{2m} \nabla_i^2 + V_{\text{nucleus}} \right] + \sum_{i<j} \frac{e^2}{|r_i - r_j|},$$

where $\hat{H}$ is the Hamiltonian operator representing the total energy of the system. The first summation $\sum_i$ runs over all electrons. For each electron i, we calculate its kinetic energy and its interaction with the nucleus. The second summation $\sum j$ represents the repulsive Coulomb interaction between all pairs of electrons, where e is the elementary charge, and $\mathrm{r_i}$ and $\mathrm{r_j}$ are the positions of two electrons.

In the above Hamiltonian, the kinetic energy of an electron is given by $\frac{\hbar^2}{2m}\nabla_i^2$, where $\hbar$ is the reduced Planck's constant and m is the mass of electron.

The interaction between each electron with nucleus is described by the potential, $V_{\text{nucleus}}(r_i) = -\frac{Ze^2}{r_i}$, where Z is the atomic number of the nucleus and r_i is the distance between the electron and the nucleus, representing the Coulomb attraction between the negatively charged electron and the positively charged nucleus.

The repulsive Coulomb interaction between electrons is modeled by $\frac{e^2}{|r_i - r_j|}$, where r_i and r_j are the positions of two electrons, accounting for the repulsion between them. However, as the number of electrons increases, the number of pair-wise interactions grows rapidly, leading to a computational explosion that makes it increasingly difficult to model large systems accurately. To address this complexity, various approximation methods have been developed to simplify the many-body problem and make it computationally feasible.

1.3 Introduction to Approximation Methods

The primary goal of introducing approximation methods is to simplify the Schrödinger equation, allowing for tractable solutions that can predict key physical properties of systems such as energy, magnetic behavior, and electronic structure. While exact solutions are typically out of reach for large systems, these approximations allow for efficient calculations and predictions, especially in the case of complex materials, molecules, and nanostructures.

1.3.1 The Hartree Approximation Method

The first major approximation method developed to simplify the many-body problem was the Hartree approximation. Building on the need to handle the complex interactions between electrons, the Hartree approximation introduced the idea of treating each electron as moving independently in an average potential created by all other electrons. This was a significant simplification compared to the full many-body problem, where each electron interacts with every other electron.

In the Hartree approximation, the total wavefunction of a system of N electrons is approximated as a product of individual single-particle wavefunctions. Hence, the total wavefunction is given by

$$\psi(r_1, r_2, \ldots, r_n) = \prod_{i=1}^{N} \psi_i(r_i),$$

where r_i represents the position (and possibly spin) of the ith electron, and $\psi_i(r_i)$ is the wavefunction for the ith electron. In this approximation, it was assumed that electron moves independently and it feels the mean-field potential of other electrons. This drastically reduces the complexity of the many-body wavefunction as it becomes a product of single-particle wavefunction rather than a multi-dimensional function of all electrons.

Mean-Field Approximation

In the Hamiltonian, the term that describes coulomb interaction between electrons is replaced by a mean-field approximation. This means that the total electron-electron repulsion is approximated by an effective potential felt by each electron, rather than considering the exact repulsion between every pair of electrons. This effective potential is the average potential felt by an electron in a system, which was created by the rest of the electrons in the same system.

The effective potential is also known as Hartree potential (V_H), and it was described by the following equation:

$$V_H(r) = \int \frac{\rho(r')}{|r - r'|} d^3r',$$

where $\rho(r')$ is the electron density at position r', defined as

$$\rho(r') = \sum_{i=1}^{N} |\psi_i(r')|^2.$$

Now the Hamiltonian in the Hartree approximation for each electron can be written as

$$\hat{H} = -\frac{\hbar^2}{2m}\nabla_i^2 + V_{\text{nucleus}}(r) + V_H(r).$$

Hence, the Schrödinger equation for each electron is given as

$$\left[-\frac{\hbar^2}{2m}\nabla_i^2 + V_{\text{nucleus}}(r) + V_H(r)\right]\psi_i(r) = \varepsilon_i\psi_i(r),$$

where $\psi_i(r)$ is the single-particle wavefunction for the ith electron and ϵ_i is its energy.

This equation can be solved self-consistently for each electron as the total electron density $\rho(r)$ is iteratively updated while the solutions for $\psi_i(r)$ are refined [3].

Limitations of Hartree Approximation

In this approximation, it is assumed that electrons are independent and each electron interacts with other electrons only through an averaged electrostatic potential created by them. Thus, the degree of freedom of the electron spin or exchange interaction of electrons was not considered. Therefore, the approximation doesn't enforce the antisymmetry of the total wavefunction for electrons, which is a fundamental requirement of quantum mechanics due to the Pauli exclusion principle. To address this issue, a new approach known as Hartree–Fock approximation was introduced.

1.3.2 Hartree–Fock Approximation

This approximation extends the Hartree method by incorporating the exchange interaction, ensuring a more realistic description of electron-electron interactions. The key innovation of Hartree–Fock is the use of the Slater determinant, a mathematical construct that automatically enforces the required antisymmetry of the wavefunction. This modification introduces an additional term, known as the exchange potential, which corrects the deficiencies of the Hartree approximation. The determinant for a many-electron system consisting N number of electrons was expressed as:

$$\psi_{HF} = \frac{1}{\sqrt{N!}} \begin{bmatrix} \psi_1(r_1) & \psi_1(r_2) & \cdots & \psi_1(r_N) \\ \psi_2(r_1) & \psi_2(r_2) & \cdots & \psi_2(r_N) \\ \vdots & \vdots & \ddots & \vdots \\ \psi_N(r_1) & \psi_N(r_2) & \cdots & \psi_N(r_N) \end{bmatrix},$$

where

- Ψ_{HF}: The Hartree–Fock wavefunction, describing the quantum state of the system.
- $\psi_i(r_i)$: The single-particle orbital of the ith electron at position r_i, representing the spatial distribution of the electron [4].

The determinant ensures that swapping any two electrons flips the sign of the wavefunction, thus creating the exchange effect. For example:

$$\Psi_{HF}(r_1, r_2, \ldots, r_N) = -\Psi_{HF}(r_2, r_1, \ldots, r_N).$$

The determinant changes sign when any two electrons are swapped, which means that the electrons with same spin are forced to stay apart, reducing repulsion. This creates the exchange interaction, which lowers the total energy of the system compared to Hartree method. Despite this improvement, Hartree–Fock (HF) approximation still has several limitations.

Limitations of Hartree–Fock Approximation

Even though HF accounts for the exchange interaction as discussed earlier, it neglects electron correlation because it is still considering the mean-field potential concept introduced in Hartree approximation method. Electron correlation refers to the dynamic interactions between electrons where they adjust their positions instantaneously in response to each other. In simple terms, when electrons are moving closer to each other, they experience repulsive forces and they avoid each other due to their like charges. This means that the position and motion of one electron is affecting the other, thus their behavior becomes correlated [1].

This limitation urged a shift toward more advanced approximation methods. As a result, researchers began to explore post-Hartree–Fock (post-HF) methods, which go beyond the mean-field approximation of HF by incorporating electron correlation effects more accurately. Techniques such as Configuration Interaction (CI), and Coupled Cluster (CC) were developed to provide a more precise description of many-electron systems. In the next section, these post-HF methods and their role in improving electronic structure calculations will be discussed briefly.

1.3.3 Configuration Interaction (CI)

As a post-HF method, CI systematically improves electronic structure calculations by expanding the wavefunction into a linear combination of multiple slater determinants.

CI includes excited-state configurations in these determinants, capturing more of the electron correlation effects. The general expression of linear combination of multiple determinants in CI approach is give as

$$|\Psi_{CI}\rangle = c_0|\Phi_0\rangle + \sum_{i,a} C_i^a|\Phi_i^a\rangle + \sum_{i<j,a<b} C_{ij}^{ab}|\Phi_{ij}^{ab}\rangle + \ldots,$$

where:

- $|\Phi_0$ is the Hartree–Fock reference configuration (ground state)
- $|\Phi_i^a$ represents an electron excitation, where an electron moves from an occupied orbital i to an unoccupied virtual orbital a. (single excitation)
- $|\Phi_{ij}^{ab}$ represents a double excitation, where two electrons move from orbitals i, j to a, b.
- The coefficients $c_0, c_i^a, c_{ij}^{ab}, \ldots$ determine how much each configuration contributes to the total wavefunction.

The goal of CI is to find the coefficients c_0, c_i^a, c_{ij}^{ab},...by solving the Schrödinger equation:

$$H|\Psi_{CI}\rangle = E_{CI}|\Psi_{CI}\rangle.$$

This leads to the CI matrix equation:

$$Hc = E_{CI}c,$$

where H is the Hamiltonian matrix and c is the vector of CI coefficients. Here, the lowest eigenvalue of H gives the ground-state CI energy and higher eigenvalues correspond to the excited-state energies. The CI coefficients reveal the contribution of different electron configurations, providing insight into the nature of electronic wavefunction [5].

Although CI is a powerful method for computing electronic excitation spectra, it is computationally infeasible for large systems due to the exponential scaling of Hamiltonian matrix size. Moreover, it suffers from the lack of size extensivity, meaning the total energy does not scale properly with system size. To overcome these issues, a method like Coupled Cluster (CC) is often preferred for large molecular systems, as it incorporates size extensivity while maintaining high accuracy.

1.3.4 Coupled Cluster (CC) Method

Unlike CI, which express the wavefunction as a linear combination of slater determinants, CC introduces an exponential ansatz for the correlated wavefunction. The CC expression is given as

$$|\Psi_{CC}\rangle = e^{\hat{T}}|\Phi_0\rangle,$$

where:

- $|\Phi_0$ is the HF reference determinant,
- $\hat{T}$ is the cluster operator, defined as

$$\hat{T} = \hat{T}_1 + \hat{T}_2 + \hat{T}_3 + \ldots.$$

Here $\hat{T}_1$ generates single excitations while $\hat{T}_2$ and $\hat{T}_3$ generate double and triple excitations respectively.

The key feature of the exponential operator $e^{\hat{T}}$ is that it naturally includes higher-order excitations through a Taylor series expansion:

$$e^{\hat{T}} = 1 + \hat{T} + \frac{\hat{T}^2}{2!} + \frac{\hat{T}^3}{3!} + \ldots.$$

This expansion leads to disconnected excitations, which are missing in CI methods but are crucial for achieving size-extensivity.

The CC method was solved using Schrödinger equation as

$$H|\Psi_{CC} = E_{CC}|\Psi_{CC}\rangle.$$

$|\Psi_{CC}$ can be replaced with the exponential term $e^{\hat{T}} \mid \Phi_0$, and hence the equation is rewritten as

$$e^{-\hat{T}}He^{\hat{T}}|\Phi_0 = E_{CC}|\Phi_0\rangle.$$

Here, $e^{-\hat{T}}He^{\hat{T}}$ transforms the Hamiltonian into a simpler form, known as the similarity-transformed Hamiltonian $(\overline{H})$. This removes the need to explicitly handle excited-state determinants.

To solve for the cluster amplitudes, a projection method is applied as below:

$$\Phi_i^a \mid \overline{H} \mid \Phi_0 = 0.$$

$$\Phi_{ij}^{ab} \mid \overline{H} \mid \Phi_0 = 0,$$

where:

- $|\Phi_i^a\rangle$ is a singly excited determinant,
- $|\Phi_{ij}^{ab}\rangle$ is a doubly excited determinant.

Since the exponential term makes the CC equations non-linear, they were solved iteratively using methods like Newton-Raphson, adjusting the cluster amplitudes step by step until the energy stabilizes (converges) [6].

Beyond CI and CC, other wavefunction-based methods have been developed to balance accuracy and efficiency. Techniques like Many-Body Perturbation Theory (MBPT) use a systematic series expansion to approximate correlation effects [7], while Multi-Reference Configuration Interaction (MRCI) extends CI to handle cases where a single reference wavefunction is insufficient [8]. These methods, while effective, still suffer from scaling challenges when applied to complex molecular systems.

Given these computational limitations, researchers have explored alternative approaches that shift the focus from explicitly solving for the many-electron wavefunction to describing the system in terms of its electron density. The electron density, which represents the probability distribution of electrons in space, provides a more computationally efficient way to determine electronic structure. Unlike wavefunction-based methods, which require tracking all electron-electron interactions explicitly, density-based methods focus on total electron distribution, significantly reducing computational complexity while still capturing essential correlation effects. In the next section, this concept is discussed in detail.

1.4 The Concept of Electronic Density

The concept of electron density is an efficient method to overcome the complexity and inaccuracies faced by the above-mentioned Schrödinger equation-based approximations.

Electron density refers to the spacial distribution of electrons, providing a simplified representation of their behavior in 3-D space. By focusing on electron density rather than individual wavefunctions, the many-body problem can be simplified significantly. This offers a more efficient and practical pathway in energy calculations. In this section, the concept and significance of electronic density will be explored.

1.4.1 Defining Electron Density

Electron density is a scalar function that gives the number of electrons in a unit volume at any point in space. Mathematically, it is defined as the sum of the absolute square of the wavefunctions of all the electrons in the system [9]:

$$\rho(r) = \sum_i |\psi_i(r)|^2,$$

where $\psi_i(r)$ is the wavefunction of the ith electron and $\rho(r)$ is the electron density at position r. the key concept here is to use total electron density of a system instead of dealing with wavefunction of each individual electron. The total electron density of a system is sufficient to describe many physical properties such as energy, charge distribution, and chemical reactivity in a much more computationally efficient way.

The role of electron density in describing many-electron systems was solidified through the Hohenberg–Kohn theorems. These theorems demonstrated that the ground state energy of a system is uniquely determined by its electron density and electron density can be used to describe all essential properties of the system.

1.4.2 Hohenberg–Kohn Theorems

First Hohenberg–Kohn theorem: Once the ground-state electron density of a system is determined, all properties of a quantum system including its total energy can be calculated without needing to reference the individual wavefunctions of the electrons.

Second Hohenberg–Kohn Theorem: The electron density that minimizes the energy functional is the true ground-state density of the system.

These theorems show that electron density is the fundamental variable in describing many-electron systems, leading to a simplification of the problem that was previously dominated by complex many-body wavefunctions. Building on Hohenberg–Kohn theorems, a powerful quantum mechanical tool known as density functional theory (DFT) was developed [10].

1.4.3 Density Functional Theory (DFT)

The Hohenberg–Kohn theorems implied the existence of a universal energy functional $E[\rho]$ of a system and the ground-state energy of it is obtained by minimizing this energy functional. It means $E_0 = minE[\rho]$ where E_0 is the true ground-state energy. While these theorems provide the conceptual basis for DFT, they do not provide the exact functional form of $E[\rho]$ which includes the kinetic energy and electron-electron interactions. Therefore, this lead to the development of a practical approximation, known as Kohn–Sham approach. Kohn and Sham introduced an auxiliary system of non-interacting electrons that reproduces the exact ground-state electron density of the real interacting system. This auxiliary system is governed by an effective potential, which ensures that the resulting electron density matches that of the true interacting system.

The Kohn–Sham Energy Functional

Kohn and Sham proposed that the total energy functional can be written as:

$$\mathrm{E}[\rho] = \mathrm{T_s}[\rho] + \mathrm{V_{nucleus}}[\rho] + \mathrm{V_H}[\rho] + \mathrm{E_{XC}}[\rho]$$

where:

- $T_s[\rho]$ is the kinetic energy of non-interacting electrons,
- $E_{XC}[\rho]$ is the exchange-correlation energy, which encapsulates all quantum many-body effects.

With the introduction of $E_{XC}[\rho]$, Kohn and Sham were able to effectively separate the many-body problem into a system of independent particles moving in an effective potential.

Kohn–Sham Equations

To determine the electron density $\rho(r)$, Kohn and Sham introduced a set of equations similar to the Schrödinger equation:

$$\left[-\frac{\hbar^2}{2m}\nabla^2 + V_{\text{eff}}(r)\right]\phi_i(r) = \epsilon_i\phi_i(r)$$

$\phi_i(r)$ represent the Kohn–Sham orbitals and ϵ_i represent their eigenvalues. The effective potential $V_{eff}(r)$ is derived from the functional components as:

$$\mathrm{V_{eff}(r) = V_{nucleus}(r) + V_H(r) + V_{xc}(r)}$$

Here the kinetic energy term $T_s[\rho]$ is implicitly included in the kinetic energy operator $\frac{\hbar^2}{2m}\nabla_i^2$. The exchange-correlation potential $V_{xc}(r)$ captures many-body effects beyond Hartree term. $V_{xc}(r)$ is defined as:

$$\mathrm{V_{xc}(r) = \frac{\delta E_{xc}[\rho]}{\delta\rho(r)}}$$

Thus, the electron density can be reconstructed from the Kohn–Sham orbitals as:

$$\rho(\mathrm{r}) = \sum_i |\psi_i(r)|^2$$

The Kohn–Sham equations provide a self-consistent framework where the total energy functional dictates the effective potential, which in turn determines the electronic density and orbitals [11].

Having established the Kohn–Sham equations as the foundation for DFT calculations, the iterative process involved in solving the Kohn–Sham equations are mentioned below:

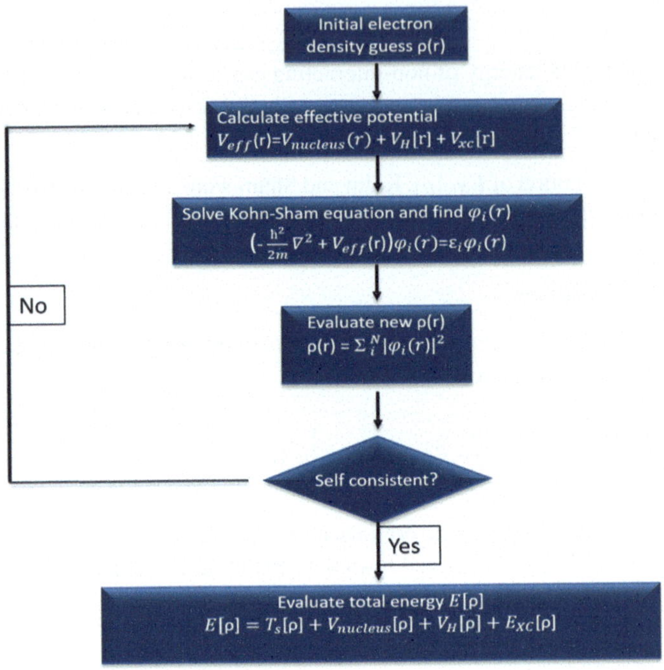

As illustrated in the flowchart, DFT process provides a self-consistent framework for determining the ground-state electron density and total energy of a system [12]. By iteratively solving the Kohn–Sham equations, updating the electron density, and checking for convergence, DFT enables accurate electronic structure calculations with relatively low computational cost compared to many-body wavefunction methods.

The total energy of the system, which is the key outcome of the process, serves as a foundation for geometry optimization. By minimizing the total energy with respect to nuclear positions, one can determine the most stable molecular structures. This process, known as structural relaxation, allows researchers to predict bond lengths, angles, and overall molecular configurations with high precision. Moreover, total energy differences between various conformations enable the identification of the most energetically favorable states, aiding in reaction mechanism studies and phase stability analysis.

DFT plays a crucial role in materials science, quantum chemistry, and condensed matter physics. It is instrumental in optimizing molecular structures, understanding reaction mechanisms, and designing molecular structural combinations with unique properties.

References

1. L.J. Bartolotti, K. Flurchick, An introduction to density functional theory. Rev. Comput. Chem. **7**, 187–216 (2007). https://doi.org/10.1002/9780470125847.ch4
2. S.E.E. Profile, M. Kozlowski The Schrödinger equation A History (2019)
3. The Hartree Approximation. Accessed 26 Mar 2025. https://www.tcm.phy.cam.ac.uk/~ajw29/thesis/node7.html
4. 9.3 The Hartree-Fock Approximation. Accessed: 26 Mar 2025. https://web1.eng.famu.fsu.edu/~dommelen/quantum/style_a/hf.html
5. X. Wu, P. Su, Very brief introduction to quantum chemistry. Quantum Chem. Age Mach. Learn., 3–25 (2023). https://doi.org/10.1016/B978-0-323-90049-2.00006-8
6. Coupled cluster — Introduction to Electronic Structure Methods. Accessed 26 Mar 2025. https://lcbc-epfl.github.io/iesm-public/Lecture/CC.html
7. H. Gould, J. Tobochnik, Introduction to many-body perturbation theory. Stat. Therm. Phys. 405–414 (2006) http://stp.clarku.edu/notes/
8. Comprehensive Inorganic Chemistry II|ScienceDirect. Accessed 26 Mar 2025. https://www.sciencedirect.com/referencework/9780080965291/comprehensive-inorganic-chemistry-ii
9. N. Salami, A. Shokri, Electronic structure of solids and molecules. Interface Sci. Technol. **32**, 325–373 (2021). https://doi.org/10.1016/B978-0-12-818806-4.00002-4
10. S. Kurth, M.A.L. Marques, E.K.U. Gross, Density-functional theory. Encycl. Condens. Matter Phys., 395–402 (2005). https://doi.org/10.1016/B0-12-369401-9/00445-9
11. G. Gottstein, M. Goerdeler, G.V.S.S. Prasad, Encyclopedia of condensed matter physics, Encycl. Condens. Matter Phys. 298–305 (2005) Accessed 26 Mar 2025. http://www.sciencedirect.com:5070/referencework/9780123694010/encyclopedia-of-condensed-matter-physics
12. T.D. Widayani, K. Wungu, S.E. Marsha, Suprijadi, Study of target recognition of MAA-based molecularly imprinted polymer (MIP) using density functional theory (DFT) computation on the interaction of Methacrylic acid (MAA)-D-glucose. J. Polym. Biopolym. Phys. Chem **5**, 10–12 (2017). https://doi.org/10.12691/JPBPC-5-1-2

Chapter 2
Fundamentals of DFT

Piumantha Samaranayake, Visal de Silva, Azeez Ahamed, Nadeesha Manohari Wickramage, and Roshan Thotagamuge

Abstract The foundations of density functional theory (DFT), a computational approach to quantum mechanical modeling, are examined in this chapter. The historical evolution of DFT is covered first, emphasizing how well it approximates the Schrödinger equation and how widely it can be used to determine the electrical and molecular characteristics of different materials. Hohenberg and Kohn's theorems from 1964 address the constraints of the Hartree and Thomas–Fermi models, which are cited as early forerunners. These theorems made electron density a fundamental variable in ground state property calculations. The Kohn–Sham (KS) method, which introduces non-interacting electrons in an effective potential to simplify the many-electron problem, is then discussed. The foundation for self-consistent computations is this framework, which incorporates exchange–correlation functionals to take quantum mechanical interactions into account. The chapter discusses the accuracy and computational cost trade-offs of several exchange–correlation functional types, including hybrid functionals, Generalized Gradient Approximation (GGA), and Local Density Approximation (LDA). The chapter firmly establishes DFT's place in contemporary computational science by highlighting the iterative nature of DFT calculations and the ongoing improvements in functional development to improve accuracy across a variety of chemical and physical systems.

Keywords Density functional theory (DFT) · Hohenberg–Kohn (H–K) · Kohn–Sham (K-S) · Self-Consistent Field (SCF) · Local Density Approximation (LDA) · Generalized gradient approximation (GGA)

P. Samaranayake (✉) · V. de Silva · A. Ahamed · N. M. Wickramage
Department of Physics, Faculty of Science, University of Ruhuna, Matara, Sri Lanka
e-mail: samaranayakepiumantha@gmail.com

R. Thotagamuge (✉)
Department of Optometry, Faculty of Allied Health Sciences, University of Sri Jayewardenepura, Nugegoda, Sri Lanka
e-mail: roshan@sjp.ac.lk

R. Thotagamuge (ed.), *Density Functional Theory*,
https://doi.org/10.1007/978-981-95-9745-1_2

2.1 History and Development of DFT

Numerous techniques have been created to roughly solve the Schrödinger [1, 2] and Dirac equations [3, 4]. In recent decades, density functional theory (DFT) has become the most widely accepted. DFT is a quick and inexpensive quantum mechanical (QM) theory that may be used to accurately calculate a wide range of solids' physical properties. This area of study includes the creation of new analytical methods aimed at creating accurate exchange–correlation functionals as well as the application of this method to forecast the electronic and molecular makeup of atoms, molecules, complexes, and solids in both gas and solution phases. The search for the exchange–correlation functional, which uses density to depict complex many-body processes inside one element formalism, is what led to DFT's success [5]. The Hartree function is a method that Hartree invented in 1927 to determine estimated wavefunctions and energy for atoms and ions. His students Fock and Slater separately proposed a self-consistent function with consideration of Pauli principles and the multi-electron wavefunction in the form of a determinant of one-particle orbitals (Slater-determinant) a few years later in order to get around the complete disregard of the anti-symmetry of the electron system [6]. The Hartree–Fock model wasn't widely used until the 1950s because of how difficult it was to calculate. The Hartree–Fock model, however, assumes that the outcome will only be an approximation of the actual outcome.

A statistical model to calculate an atom's energy by approximating its electron distribution was put forth by Thomas and Fermi in 1927 [7, 8], the same year that the Hartree function was introduced. The kinetic energy of an atom was described using the electron density functional, and the atom energy was calculated by adding the two traditional terms for nuclear-electron and electron–electron interactions, both of which can be expressed in terms of electron density. It initially lacked the atom's exchange energy, which is a result of the Pauli principle and expressed in Hartree–Fock theory. Dirac added an exchange energy functional term in 1928. The Thomas–Fermi model is a crucial first step, but its applications are currently very limited because it is inaccurate for the majority of them. The exchange energy has the largest source of error due to the complete neglect of the electron correlation effect, followed by the representation of the kinetic energy functional term, which is only an approximation. Nevertheless, it is a forerunner of the density functional theory (DFT). Hohenberg and Kohn laid the groundwork for the DFT Mansion Firm in 1964 when they released a study. The fundamental idea behind DFT is to replace the complex and therefore challenging-to-calculate many-electron wavefunction, which has 3N variables (where N is the number of electrons and each electron has three spatial variables), with the electron density's functional, which only has three variables and is the function of another function that maps a number to a function. Therefore, we just deal with three variables in the new system, which is much easier to manage than the enormous number of 3N variables. Kohn and Hohenberg put out their initial theorem, which emphasizes that the ground state energy is a function of electron density and that it depends exclusively on it. Their second theorem demonstrated that ground state energy may be derived by minimizing the system's energy in accordance with the electron density [9]. It is necessary to acknowledge that H–K theorems simply confirm the existence of one-to-one mapping links between electron density functional and system attributes; they do not specify the nature of these interactions. Therefore, the Kohn–Sham method is the most widely utilized approach rather than "minimizing the system energy." In 1965, just a year after Hohenberg and Kohn's seminal publication, Kohn and Sham released a study in which they reduced the multi-electron problem to a problem of non-interaction electrons in an effective potential. The effects of the Coulomb interactions between the electrons, such as the exchange and correlation interactions, are included in this potential together with the external potential. The challenge with KS-DFT is handling the exchange and correlation interaction. As of right now, the exchange and correlation energy cannot be solved rigorously. Nonetheless, the Local-Density Approximation (LDA) is the most straightforward approximation. The foundation of LDA is the use of the uniform electron gas model to obtain the correlation energy from fits to the uniform electron gas and the exchange energy, the precise value of which may be obtained using the Thomas–Fermi model. The kinetic energy functional of this system is precisely known by converting the problem into a non-interacting system in an effective potential, where the wavefunction is readily represented by a Slater determinant of orbitals. However, nothing is known about the exchange–correlation component of the total energy functional.

The energy of a non-interacting reference system is computed using Kohn–Sham DFT. The difference to the actual system is approximated using exchange and correlation functionals. The electron density alone determines the most basic exchange and correlation functionals. More precise computations were made possible by the creation of exchange and correlation functionals that additionally incorporated the gradient (GGA functionals) [10–12] and second derivatives (meta-GGA functionals) [13–15] of the electron density. Higher precision is generally achieved using GGA and meta-GGA computations at a very small additional computational cost. The creation of hybrid functionals made DFT computations considerably more appealing and accurate [16–20].

Hybrid functionals, however, also markedly raised the computational cost. Higher accuracy for atoms and molecules was obtained by mixing a global percentage of Hartree–Fock exchange, and range-separation using a screened Coulomb potential made it possible to apply hybrid functionals to solids and surfaces [21–23]. Reaction barrier height calculations were more precisely performed using long-range-corrected hybrid functionals [24, 25]. The addition of a variable proportion of Hartree–Fock exchange, as accomplished via local hybrid functionals, further improved the accuracy of DFT computations [26–30]. The accuracy was further improved by combining the ideas of range-separated hybrids and local hybrids [31–33].

Furthermore, the original correlation functional was combined with correlation derived from wave function techniques (MP2, RPA, and coupled-cluster). Based on

Runge and Gross's work, time-dependent DFT has established itself as a reliable technique for handling electronically excited states.

2.2 Hohenberg and Kohn Theorems

In 1964, Hohenberg and Kohn [34] proposed two seemingly straightforward theorems that made it possible to apply DFT.

2.2.1 The First H–K Theorem

In their first theorem, Hohenberg and Kohn demonstrated that an electron density that depends on just three spatial coordinates (r has three variables) is the sole way to identify the ground state features of a many-electron system. They arrived at their first theorem by taking into consideration a group of an arbitrary number of electrons traveling under the influence of the mutual Coulomb repulsion and an external potential v(r) while enclosed in a big box [35]. The form of the Hamiltonian is,

$$H = T + V + U, \tag{2.1}$$

Where

$$T = \frac{1}{2}\int \nabla\psi^*(r)\int \nabla\psi(r)dr \tag{2.2}$$

$$V = \int v(r)\psi^*(r)\psi(r)dr \tag{2.3}$$

$$U = \frac{1}{2}\int \frac{1}{|r - r'|}\psi^*(r)\psi^*(r')\psi(r)drdr', \tag{2.4}$$

where U is the electron–electron interaction energy, V is the external field's N-electron potential energy, and T is the N-electron kinetic energy. Only systems with a nondegenerate ground state are addressed here. It takes Time Depend DFT to handle degenerated states when there is a magnetic field present.

Next, observe the ground state electron density ψ via

$$n(r) \equiv \left(\Psi, \psi^*(r)\psi(r)\Psi\right). \tag{2.5}$$

Now the logic is: from the expression above, n(r) is clearly a functional of v(r), because wavefunction ψ(r) can be decided by v(r). If it can be proved that v(r) is a unique functional of n(r), and because v(r) can ψ(r), can fixes H, then if provided n(r), all the properties of ground state are unique functionals of n(r).

Assume another potential v'(r), with ground state Ψ' can give the same density n(r). Ψ' and Ψ can not be equal to each other because they satisfy different Shrödinger equations, so they have different Hamiltonians H and H′ and different ground state energies E and E′ [35] and they must meet.

$$E' = \left(\Psi', \mathrm{H}'\Psi'\right) < \left(\Psi, \mathrm{H}'\Psi\right) = (\Psi, (\mathrm{H} + \mathrm{V}' - \mathrm{V})\Psi).$$

So that

$$E' < E + \int \left[v'(r) - v(r)\right] n(r) dr. \tag{2.6}$$

To obtain the same result, simply switch the primed and unprimed values and repeat the computation:

$$E < E' + \int \left[v'(r) - v(r)\right] n(r) dr. \tag{2.7}$$

By adding the Eqs. (2.6) and (2.7) can get

$$E' + E < E + E'. \tag{2.8}$$

This is only possible when $E = E'$. This indicates that v(r) is a unique functional on n(r), which in turn indicates that the complete many-particle ground state is a unique functional on n(r). It also suggests that v'(r) cannot originate from the same n(r).

2.2.2 The Second H–K Theorem

Hohenberg and Kohn define an energy functional in terms of the system's electron density in their second theorem. They further demonstrate that ground state energy may be obtained by taking the minimum of the energy functional based on the electron density.

As previously stated, all of the ground state's characteristics are now functionals of electron density, so define

$$F[n(r)] \equiv (\Psi, (T + U)\Psi). \tag{2.9}$$

In this case, F[n] is a universal functional that holds true for any number of particles and any external potential; however, it becomes non-universal if it incorporates V, as V is dependent on the system [35]. Next, given v(r), an external potential, define

$$E_V[n] \equiv \int v(r)n(r)dr + F[n]. \tag{2.10}$$

This is the ground state energy E for the proper n(r).

The reasoning for this is that the ground state electron density, n(r), will provide the functional with the lowest value, and no other electron density will provide an energy greater than the ground state energy. By doing this, first there should be an restrict condition

$$N[n] \equiv \int n(r)dr = N. \tag{2.11}$$

Given that the system's electron count is constant at N, this is evidently true [35].

From the Eq. (2.1) it is easy to get for a system Ψ', the energy functional is

$$\varepsilon_v[\Psi'] \equiv (\Psi', V\Psi') + (\Psi', (T + U)\Psi'). \tag{2.12}$$

It has a minimum at the correct ground state Ψ, and all other state Ψ' other than the ground state will provide the larger value. Let Ψ' be the ground state of the system with a different external potential v′(r), using the Eq. (2.12) and substitute (2.9) into the later part of (2.12) can get

$$\varepsilon_v[\Psi'] = \int v(r)n'(r)dr + F[n'] > \varepsilon_v[\Psi] = \int v(r)n(r)dr + F[n] \tag{2.13}$$

Remember here Ψ denotes the ground state with external potential with v(r), and Ψ' is other state which now is set to represent the ground state with external potential v′(r). And we are only interested in v(r), not v′(r). Now the point is proved.

Finding the correct equation for F[n] is quite difficult, but once it is, we can obtain the ground state energy by inserting it into (2.10), calculating the derivative of it, and determining the lowest value of the functional. However, the KS-DFT is typically a better method.

2.3 Kohn–Sham Equations

The Kohn–Sham (KS) equations are important to density functional theory (DFT) because users allow to solve for the ground-state electron density in interacting many-body systems. The fundamental benefit of employing the KS equations is that they convert a complex interacting system to a non-interacting one, making it computationally helpful to investigate as large systems like molecules, solids, and nanomaterials [36].

The KS equations are a set of Schrödinger-like equations for single particles that describe the behavior of electrons in an effective potential. The key concept is that the interacting many-body problem is translated into an equivalent non-interacting problem whose solutions (the Kohn–Sham orbitals) produce the identical ground-state electron density [37].

2.3.1 Basic Form of the Kohn–Sham Equations

The KS equations are expressed as

$$\left(-\frac{\hbar^2}{2m}\nabla^2 + V_{eff}(r)\right)\psi_i(r) =_i \psi_i(r), \tag{2.14}$$

where

$\psi_i(r)$ are the Kohn–Sham orbitals, representing the wavefunctions of non-interacting electrons.

ϵ_i are the energy eigenvalues corresponding to these orbitals.

$V_{eff}(r)$ is the effective potential, which combines the external potential (due to the nuclei), the Hartree potential (due to electron–electron repulsion), and the exchange–correlation potential (which accounts for quantum mechanical effects in electron–electron interactions).

The effective potential is the most crucial component of DFT. This takes consideration of all of the effects of electron–electron interactions, which are not explicitly addressed in the Schrödinger equation for non-interacting particles [38].

2.3.2 Electron Density and the Role of Kohn–Sham Orbitals

The electron density $\rho(r)$ is a fundamental quantity in DFT, which is related to the Kohn–Sham orbitals. The total electron density is given by:

$$\rho(r) = 2\sum|\psi_i(r)|^2, \tag{2.15}$$

where the sum runs over all occupied orbitals (those with energy $\epsilon_i \leq 0$).

The factor of two accounts for the fact that each orbital may contain two electrons with opposite spins (since we typically consider spin-paired electrons). The electron density ρ(r) is the main variable in the DFT formalism, determining the system's total energy and other electronic properties [39].

2.3.3 Effective Potential $V_{eff}(r)$

In the Kohn–Sham equations, the effective potential $V_{eff}(r)$ consists of several terms:

- **External Potential**: $V_{ext}(r)$ describes the interaction of electrons with the external potential, typically the Coulomb potential from the nuclei.
- **Hartree Potential**: $V_H(r)$ describes the classical electrostatic repulsion between electrons, given by:

$$V_H(r) = \int \frac{\rho(r')}{|r - r'|} dr' \tag{2.16}$$

- **Exchange–Correlation Potential**: The most problematic aspect of DFT is $V_{xc}(r)$, which encompasses all quantum mechanical effects resulting from the Pauli exclusion principle and electron correlation. It is the key factor that enables DFT to move beyond classical electrostatics [40].

Now, the effective potential can be written as:

$$V_{eff}(r) = V_{ext}(r) + V_H(r) + V_{xc}(r) \tag{2.17}$$

This potential is self-consistently updated during the iterative process of solving the KS equations [41].

2.3.4 Self-consistent Field (SCF) Procedure

1. **Initial Guess**:

The trial electron density, $\rho(r)$, is assumed. This can be done using a uniform electron gas, a superposition of atomic densities, or a previously converged density from a similar system.

This guess has significance since it initiates the iterative process.

2. **Calculate Effective Potential**:

The effective potential $V_{eff}(r)$ is computed. This one includes the external potential, the Hartree potential, and the exchange–correlation potential.

3. **Solve the Kohn–Sham Equations**:

Kohn–Sham (KS) equations are solved for single-particle wavefunctions, $\psi_i(r)$. These equations explain how non-interacting electrons move within the effective potential calculated in the previous stage.

4. **Evaluate the Electron Density and Total Energy**:

The electron density $\rho(r)$ is updated using the Kohn–Sham orbitals. This density is computed as follows:

$$\rho(r) = \sum_i f_i |\psi_i(r)|^2, \tag{2.18}$$

where f_i represents the occupation of each orbital (typically 2 for spin-paired orbitals).

The total energy is also evaluated using the updated electron density.

5. **Converged**?

A convergence check is run to determine whether the updated electron density matches the previous one within a predefined threshold (e.g., $10^{-6} atomicunits$). If the difference in density between two successive iterations is less than the threshold, convergence is considered accomplished.

$$\Delta\rho = \|\rho_{new} - \rho_{old}\| < threshold \tag{2.19}$$

If the operation does not converge, it returns to step 2 and updates the potential before beginning a new iteration.

6. **Output Quantities**:

Once convergence is achieved, the SCF operation is finished. The final results include the converging electron density, Kohn–Sham orbitals, total energy, and any other needed physical parameters (e.g., forces, dipole moments, etc.) [42, 43].

2.4 Exchange–Correlation Function

The exchange–correlation function is essential in the Kohn–Sham (KS) formulation of DFT. It represents all of the complex quantum mechanical interactions between electrons that classical electrostatics cannot account for. These interactions are critical for effectively representing electron behavior in real systems [44].

Because the exact form of the exchange–correlation function is uncertain, a variety of approximations are used in practice. We'll look at the theoretical foundations, several sorts of approximations, and how the exchange–correlation functional affects DFT computations.

2.4.1 Theoretical Background of the Exchange–Correlation Functional

In DFT, the total energy of a system is a function of its electron density, $\rho(r)$. The energy functional is separated into many components, with the exchange–correlation energy, E_{xc} being the part that cannot be accounted for by the classical electrostatic interaction (Hartree term) [45]. The total energy in DFT can be calculated as:

$$E[\rho] = T_s[\rho] + E_{ext}[\rho] + E_H[\rho] + E_{xc}[\rho], \tag{2.20}$$

where:

$T_s[\rho]$ is the kinetic energy of the non-interacting system of electrons.

$E_{ext}[\rho]$ is the energy from the external potential (nuclei).

$E_H[\rho]$ is the Hartree energy (classical electron–electron repulsion).

$E_{xc}[\rho]$ is the exchange–correlation energy [46].

The exchange–correlation functional, $E_{xc}[\rho]$, captures the quantum mechanical consequences of electron–electron interactions, including:

Exchange effects: The Pauli exclusion principle results electrons with the same spin to avoid each other.

Correlation effects: These occur as a result of electrons' dynamic interactions, such as the tendency to "correlate" their motion in order to reduce repulsion [47].

The exchange–correlation energy is expressed as an integral over the electron density:

$$E_{xc}[\rho] = \int \rho(r)\epsilon_{xc}(\rho(r))dr \tag{2.21}$$

where $\epsilon_{xc}(\rho(r))$ is the exchange–correlation energy per electron at the point r.

Since the exact form of $E_{xc}[\rho]$ is not known, approximations are made.

2.4.2 Types of Exchange–Correlation Functionals

There are many types of approximations for the exchange–correlation functional in DFT. Each approximation has advantages and disadvantages, depending on the system under investigation and the level of accuracy desired.

Local Density Approximation (LDA)

The LDA assumes that the exchange–correlation energy at any point r depends only on the local electron density $\rho(r)$, ignoring the spatial variation of the density [48].

$$E_{xc}^{LDA}[\rho] = \int \rho(r)\epsilon_{xc}^{LDA}(\rho(r))dr \tag{2.22}$$

The LDA works well for systems with nearly uniform electron densities, such as metals or simple homogeneous system, However, LDA often fails for inhomogeneous systems, such as molecules or solids with complex electronic environments, where the electron density varies significantly from point to point.

Generalized Gradient Approximation (GGA)

The GGA improves upon the LDA by including not only the local electron density but also the gradient of HF the density $\nabla\rho(r)$, allowing the functional to account for the spatial variation of the density.

$$E_{xc}^{GGA}[\rho] = \int \rho(r)\epsilon_{xc}^{GGA}(\rho(r), \nabla\rho(r))dr \tag{2.23}$$

GGA provides a more accurate treatment of inhomogeneous systems, such as molecules and surfaces, where the electron density changes spatially. While GGA improves upon LDA for many systems, it still does not accurately capture the exchange–correlation energy in systems with strong electron correlation (e.g., transition metal complexes or systems with a large degree of electron localization).

Examples of commonly used GGA Functionals: PBE (Perdew–Burke–Ernzerhof), BYLP (Becke–Lee–Yang–Parr) [49].

Hybrid Functionals

Hybrid Functionals combine elements from both Hartree–Fock theory and DFT. Specially, they incorporate a portion of the exact exchange from Hartree–Fock theory along with a GGA exchange–correlation functional.

The general form for hybrid functional is:

$$E_{xc}^{hybrid}[\rho] = \int \rho(r)[a\epsilon_{HF}(\rho(r)) + (1-a)\epsilon_{xc}^{GGA}(\rho(r), \nabla\rho(r))]dr \tag{2.24}$$

where $\epsilon_{HF}(\rho)$ is the exchange energy from Hartree–Fock theory, and a is a mixing parameter that controls the amount of exact exchange incorporated.

Hybrid Functionals are more accurate than LDA and GGA for many molecular systems, especially for properties like reaction energies, molecular geometries, and bond dissociation energies. But, Hybrid functionals are computationally more expensive than pure LDA or GGA functionals, and they may not always be suitable for large systems.

Examples of commonly used hybrid functionals: B3LYP (Becke 3-parameter Lee–Yang–Parr), PBE0 (a hybrid version of the PBE functional) [50].

Meta-GGA Functionals

The Meta-GGA includes higher-order derivatives of the electron density, such as the Laplacian of the density $\nabla^2\rho(r)$, in addition to the density and its gradient.

$$E_{xc}^{meta-GGA}[\rho] = \int \rho(r)\epsilon_{xc}^{meta-GGA}(\rho(r), \nabla\rho(r), \nabla^2\rho(r))]dr \tag{2.25}$$

Meta-GGA functionals offer more flexibility and are able to provide more accurate results for a wider range of systems, including those with significant electron correlation. Meta-GGA functionals are more computationally demanding and may not always be necessary for simpler systems.

An example of a meta-GGA functional: SCAN (Strongly Constrained and Appropriately Normed) [51].

Range-Separated Hybrid Functionals

Range-Separated Hybrid Functionals split the exchange interaction into short-range and long-range components. This approach is particularly useful for systems where charge transfer is important, such as in molecules with large electron density differences.

The functional has the form:

$$E_{xc}^{range-separation}[\rho] = \int \rho(r)[\epsilon_{short}(\rho(r)) + \epsilon_{long}(\rho(r))]dr \quad (2.26)$$

These functionals are highly effective for studying systems with charge transfer, such as excited states, intermolecular interactions, or van der Waals forces. Range-separated hybrids are computationally expensive and require careful tuning of the range-separation parameter.

Examples of range-separated hybrid functionals: CAM-B3LYP (a range-separated version of B3LYP), HSE06 (a range-separated version of PBE).

The exchange–correlation functional is an important aspect of the density functional theory, accounting the complex interactions between electrons. A number of different approximations exist, each with advantages and disadvantages. This means that the search for an ever-improved form of functionals will be pursued in ongoing research within DFT. Understanding such functionals is important for choosing an appropriate functional for any given system to ensure the accuracy and efficiency of the DFT calculation.

References

1. Quantisation as an eigenvalue problem
2. E. Schrodinger, Physical review an undulatory theory of the mechanics of atoms and molecules
3. Pauli, Z.F. Physik, *The quantum theory of the electron* (1927). https://royalsocietypublishing.org/
4. S. John, *The quantum theory of the part II*. https://royalsocietypublishing.org/
5. M.A. Iqbal, N. Ashraf, W. Shahid, D. Afzal, R. Ahmad, *Fundamentals of density functional theory: recent developments, challenges and future horizons*. www.intechopen.com
6. Hartree-Fock method
7. L.H. Thomas, The calculation of atomic fields. Math. Proc. Cambridge Philos. Soc. **23**(5), 542–548 (1927). https://doi.org/10.1017/S0305004100011683
8. F. in Rom, Eine statistische Methode zur Bestimmung einiger Eigenschaften des Atoms und ihre Anwendung auf die Theorie des periodischen Systems der Elemente

9. L. Wilk A N, D.M. Nusair, Accurate spin-dependent electron liquid correlation energies for local spin density calculations: a critical analysis1 (1980)
10. J.P. Perdew, K. Burke, M. Ernzerhof, Generalized gradient approximation made simple (1996)
11. C. Lee, E. Yang, R.G. Parr, Development of the Colic-Salvetti correlation-energy formula into a functional of the electron density
12. J.P. Perdew, Density-functional approximation for the correlation energy of the inhomogeneous electron gas (1986)
13. J.P. Perdew, S. Kurth, A. Zupan, P. Blaha, Accurate density functional with correct formal properties: a step beyond the generalized gradient approximation (1999)
14. Y. Zhao, D.G. Truhlar, A new local density functional for main-group thermochemistry, transition metal bonding, thermochemical kinetics, and noncovalent interactions. J. Chem. Phys. **125**(19) (2006). https://doi.org/10.1063/1.2370993
15. J. Tao, J.P. Perdew, V.N. Staroverov, G.E. Scuseria, Climbing the density functional ladder: nonempirical meta–generalized gradient approximation designed for molecules and solids. Phys. Rev. Lett. **91**(14) (2003). https://doi.org/10.1103/PhysRevLett.91.146401
16. A.D. Becke, Density-functional thermochemistry. III. The role of exact exchange. J. Chem. Phys. **98**(7), 5648–5652 (1993). https://doi.org/10.1063/1.464913
17. C. Adamo, V. Barone, Toward reliable density functional methods without adjustable parameters: the PBE0 model. J. Chem. Phys. **110**(13), 6158–6170 (1999). https://doi.org/10.1063/1.478522
18. A.D. Becke, A new mixing of Hartree-Fock and local density-functional theories. J. Chem. Phys. **98**(2), 1372–1377 (1993). https://doi.org/10.1063/1.464304
19. Y. Zhao, N.E. Schultz, D.G. Truhlar, Design of density functionals by combining the method of constraint satisfaction with parametrization for thermochemistry, thermochemical kinetics, and noncovalent interactions. J. Chem. Theory Comput. **2**(2), 364–382 (2006). https://doi.org/10.1021/ct0502763
20. Y. Zhao, N.E. Schultz, D.G. Truhlar, Exchange-correlation functional with broad accuracy for metallic and nonmetallic compounds, kinetics, and noncovalent interactions. J. Chem. Phys. **123**(16) (2005). https://doi.org/10.1063/1.2126975
21. J. Heyd, G.E. Scuseria, M. Ernzerhof, Hybrid functionals based on a screened Coulomb potential. J. Chem. Phys. **118**(18), 8207–8215 (2003). https://doi.org/10.1063/1.1564060
22. T. Yanai, D.P. Tew, N.C. Handy, A new hybrid exchange-correlation functional using the Coulomb-attenuating method (CAM-B3LYP). Chem. Phys. Lett. **393**(1–3), 51–57 (2004). https://doi.org/10.1016/j.cplett.2004.06.011
23. J. Heyd, G.E. Scuseria, M. Ernzerhof, Erratum: hybrid functionals based on a screened Coulomb potential. J. Chem. Phys. **118**(8207) (2003). https://doi.org/10.1063/1.2204597
24. J. Da Chai, M. Head-Gordon, Systematic optimization of long-range corrected hybrid density functionals. J. Chem. Phys. **128**(8) (2008). https://doi.org/10.1063/1.2834918
25. O.A. Vydrov, G.E. Scuseria, Assessment of a long-range corrected hybrid functional. J. Chem. Phys. **125**(23) (2006). https://doi.org/10.1063/1.2409292
26. A.V. Arbuznikov, M. Kaupp, What can we learn from the adiabatic connection formalism about local hybrid functionals? J. Chem. Phys. **128**(21) (2008). https://doi.org/10.1063/1.2920196
27. R. Haunschild, B.G. Janesko, G.E. Scuseria, Local hybrids as a perturbation to global hybrid functionals. J. Chem. Phys. **131**(15) (2009). https://doi.org/10.1063/1.3247288
28. H. Bahmann, A. Rodenberg, A.V. Arbuznikov, M. Kaupp, A thermochemically competitive local hybrid functional without gradient corrections. J. Chem. Phys. **126**(1) (2007). https://doi.org/10.1063/1.2429058.
29. B.G. Janesko, G.E. Scuseria, Local hybrid functionals based on density matrix products. J. Chem. Phys. **127**(16) (2007). https://doi.org/10.1063/1.2784406
30. J. Jaramillo, G.E. Scuseria, M. Ernzerhof, Local hybrid functionals. J. Chem. Phys. **118**(3), 1068–1073 (2003). https://doi.org/10.1063/1.1528936
31. R. Haunschild, G.E. Scuseria, Range-separated local hybrids. J. Chem. Phys. **132**(22) (2010). https://doi.org/10.1063/1.3451078

32. T.M. Henderson, B.G. Janesko, G.E. Scuseria, A. Savin, Locally range-separated hybrids as linear combinations of range-separated local hybrids. Int. J. Quantum Chem. **109**(9), 2023–2032 (2009). https://doi.org/10.1002/qua.22049
33. A.V. Arbuznikov, M. Kaupp, Importance of the correlation contribution for local hybrid functionals: range separation and self-interaction corrections. J. Chem. Phys. **136**(1) (2012). https://doi.org/10.1063/1.3672080
34. T.I. Gilbert, Hohenberg-Kohn theorem for nonlocal & externai yotentiais
35. I.E. Gaa, P. Hohenbergt Ecole, X. Superzeure, F. I'aris, W. Konnt, Physical review
36. K. Burke, L.O. Wagner, DFT in a nutshell. Int. J. Quantum Chem. **112**(10), 1208–1218 (2012). https://doi.org/10.1002/qua.24259
37. G. Kresse, J. Furthmüller, Efficient iterative schemes for ab initio total-energy calculations using a plane-wave basis set. Phys. Rev. B **54**(16), 11169–11186 (1996). https://doi.org/10.1103/PhysRevB.54.11169
38. J. Lu, Numerical methods for Kohn-Sham density functional theory. Acta Numer **28**, 405–539 (2019)
39. K.T. Schütt, M. Gastegger, A. Tkatchenko, K.R. Müller, R.J. Maurer, Unifying machine learning and quantum chemistry with a deep neural network for molecular wavefunctions. Nat. Commun. **10**(1) (2019). https://doi.org/10.1038/s41467-019-12875-2
40. J.P. Perdew et al., Restoring the density-gradient expansion for exchange in solids and surfaces. Phys. Rev. Lett. **100**(13) (2008). https://doi.org/10.1103/PhysRevLett.100.136406
41. A.D. Becke, Perspective: fifty years of density-functional theory in chemical physics. J. Chem. Phys. **140**(18) (2014). https://doi.org/10.1063/1.4869598
42. J. Wang, E.J. Baerends, Self-consistent-field method for correlated many-electron systems with an entropic cumulant energy. Phys. Rev. Lett. **128**(1) (2022). https://doi.org/10.1103/PhysRevLett.128.013001
43. F. Song, J. Feng, *NeuralSCF: neural network self-consistent fields for density functional theory* (2024). http://arxiv.org/abs/2406.15873
44. W. Kohn, L.J. Sham, Self-consistent equations including exchange and correlation effects. Phys. Rev. **140**(4A), A1133 (1965). https://doi.org/10.1103/PhysRev.140.A1133
45. R.G. Parr, W. Yang, *Density-Functional Theory of Atoms and Molecules* (Oxford University Press, 1989)
46. K. Burke, Perspective on density functional theory. J. Chem. Phys. **136**(15), 150901 (2012). https://doi.org/10.1063/1.4704546
47. J.P. Perdew, K. Burke, M. Ernzerhof, Generalized gradient approximation made simple. Phys. Rev. Lett. **77**(18), 3865 (1996). https://doi.org/10.1103/physrevlett.77.3865
48. P. Hohenberg, W. Kohn, Inhomogeneous electron gas. Phys. Rev. **136**(3B), B864 (1964)
49. J.P. Perdew et al., Restoring the density-gradient expansion for exchange in solids and surfaces. Phys. Rev. Lett. **100**(13), 136406 (2008). https://doi.org/10.1103/PhysRevLett.100.136406
50. C. Adamo, V. Barone, Toward reliable density functional methods without adjustable parameters: the PBE0 model. J. Chem. Phys. **110**(13), 6158–6170 (1999). https://doi.org/10.1063/1.478522
51. J. Sun, A. Ruzsinszky, J.P. Perdew, Strongly constrained and appropriately normed semilocal density functional. Phys. Rev. Lett. **115**(3), 036402 (2015). https://doi.org/10.1103/PhysRevLett.115.036402

Chapter 3
Computational Methods and Algorithms

Vivek Pandey and Sudhir K. Pandey

Abstract Density functional theory (DFT) has proven to be an efficient tool to estimate the material properties with acceptable accuracy and thus, revolutionized the research in computational materials science. The efficiency and accuracy of DFT in exploring materials' properties depend on the basis sets used in solving the Kohn-Sham equation. Some well-known basis sets that are generally suggested for solving Kohn-Sham equation include (i) Plane waves (ii) Augmented plane wave (APW), (iii) Linearized augmented-plane-wave (LAPW), (iv) LAPW + Local Orbitals (LOs), (v) APW + local orbitals (los), (vi) Muffin-Tin Orbitals (MTO), and (vii) Linear Muffin-Tin Orbital (LMTO), etc. Each of them has their own pros and cons. Presently, the popular and efficiently employed basis set out of these are APW+lo (in DFT packages like WIEN2k and Elk), LMTO (in packages like LmtART, RSPT, and Questal), and plane waves basis set (in Abinit and Quantum Expresso). Various DFT packages use these basis sets to self-consistently solve the Kohn-Sham equation to obtain the ground state charge density, eigenstates, and eigenvalues. While solving the Kohn-Sham equation, usually the core states are not considered as they have negligible effect on most of the transport and related properties. For considering the valence states, the regions around an atom in solid is divided into two parts (I) spherical *muffin-tin* (R_{MT}) part around the nucleus of atoms where the states have steep variation and (II) interstitial regions between the neighboring spheres where there is relatively smoother variation of states of atoms forming the solid. Different types of functions are used to solve the Kohn-Sham equation in these regions. Based on this approach basis sets like APW, LAPW, and LMTO are designed. These basis sets intrinsically considers the spherical symmetric approximation of potential in the R_{MT} regions. However, these basis sets are sometimes modified to removes the spherical symmetry approximation, thus making the method more accurate [1].

V. Pandey (✉)
School of Physical Sciences, Indian Institute of Technology Mandi, Kamand, India
e-mail: vivek6422763@gmail.com

S. K. Pandey
School of Mechanical and Materials Engineering, Indian Institute of Technology Mandi, Kamand, India
e-mail: sudhir@iitmandi.ac.in

R. Thotagamuge (ed.), *Density Functional Theory*,
https://doi.org/10.1007/978-981-95-9745-1_3

This improved method is referred as *full-potential* method. There is yet another improvement in DFT implementation, where not only the valence states/electrons are treated self-consistently but also the core states/electrons. This approach of additional self-consistent treatment of DFT is the most accurate approach which is popularly referred as *all-electrons* method. Apart from these methods, there is another popular approach called Pseudopotentials method. This uses the fact that states in the regions closer to the nucleus are almost shielded from the outer portions, and act like the one in free atoms. Thus, they don't affect the regions of atoms in solids where bonding happens. This encourages to approximate the core regions by a smooth varying functions called as Pseudopotentials. Thus, the method does not explicitly treat the regions closer to the nucleus while solving the Kohn-Sham equation. This increases its computational efficiency while compromising its overall accuracy to some extent when compared with full potential approach. Present chapter will deal with these aspects along with the method to obtain converged ground state using DFT approach. The chapter will also shed light on aspects that must be taken care of while setting up any DFT calculations.

3.1 Introduction

Density Functional Theory (DFT) has emerged as one of the most powerful and versatile computational tools in modern materials science, chemistry, and physics. Rooted in the foundational Hohenberg-Kohn theorems and Kohn-Sham formalism, DFT provides a framework to solve the many-electron Schrödinger equation with remarkable accuracy and efficiency. Its ability to bridge quantum mechanics and materials' properties has transformed our understanding of a wide range of systems, from the electronic structure of molecules to the macroscopic behavior of solids.

Central to the success of DFT are its computational methods and algorithms, which translate abstract mathematical principles into practical tools for real-world applications. These methods tackle complex problems such as solving the Kohn-Sham equations, optimizing atomic geometries, and evaluating material properties under varying conditions. The choice of basis sets, the treatment of exchange-correlation functionals, and the optimization of numerical techniques play pivotal roles in balancing accuracy and computational efficiency.

This chapter delves into the computational underpinnings of DFT, offering a detailed exploration of the algorithms and strategies that drive its success. It covers essential topics, including various types of basis sets suggested for solving the Kohn-Sham equations (e.g., plane waves, APW, LAPW, LMTO), pseudopotentials, and the full-potential approach. Then, it discusses various convergence criteria used for obtaining sensible and useful results for any solid using DFT. Lastly, it discusses about some useful practical tips for setting up efficient DFT calculations. By providing a comprehensive overview, this chapter aims to equip readers with a deeper understanding of the computational frameworks that underpin DFT, enabling them to make informed choices and effectively apply these methods in their research.

3.2 What Are Basis Sets and Why Are They Needed?

In the context of crystalline solid, one is interested in knowing the eigenstates (and eigenvalues) of any given material. As Kohn-Sham equation represents a system of one electron in a periodic potential, it is required to find out all the available states (Bloch eigenstates) of the electron. There have been an uncountable discovery of crystalline material till date, and still the search is going on. Ideally, there can be infinite number of crystalline solids comprising various combinations of elements. In such scenario, there can be infinite possible types of DFT models (one electron in a periodic potential) with unique potential associated with each material. Thus, the Bloch states for such a model will be an infinite dimensional object. Basis sets are finite dimensional set of functions that are used to approximate the Bloch states with acceptable accuracy. They help in numerical representation of Hamiltonian in the matrix form and to evaluate its eigenvectors and eigenvalues. This discussion highlights the importance of basis sets for the implementation of DFT in exploring various properties of materials. In further sections, we shall discuss various types of basis set that are used (or suggested to be used) in DFT, in pedagogical manner.

3.3 Plane Waves as Basis Set

The Kohn-Sham equation in atomic units is defined as [2]-

$$-\frac{1}{2}\nabla^2\phi_i(\mathbf{r}) + V_{\text{tot}}(\mathbf{r})\phi_i(\mathbf{r}) = \varepsilon_i\phi_i(\mathbf{r}) \tag{3.1}$$

where V_{tot} contains the positive ionic (nucleus + core electrons) potential V_{n}, Hartree potential of the electrons V_{H}, and the exchange and correlation potential V_{xc}. The V_{H} is defined in terms of electronic charge density $n(\mathbf{r})$ as

$$\nabla^2 V_{\text{H}}(\mathbf{r}) = -4\pi n(\mathbf{r}) \tag{3.2}$$

where $n(\mathbf{r})$ can be obtained from the Kohn-Sham wave functions from the following relation

$$n(\mathbf{r}) = \sum_i |\phi_i(\mathbf{r})|^2 . \tag{3.3}$$

For a system with M electrons, i will run from 1-M with the $\phi_i(\mathbf{r})$ corresponding to the first N smallest eigenvalues ε_i. In the Born-Oppenheimer approximation, the positive ions are static and periodically arranged in the solids. Thus, the potential due to them (V_n) will be periodic. Moreover, the Hartree potential (V_H) from the mean-field approximation and the exchange-correlation potential (V_{xc}) add to V_n in such a way that the net potential still remains periodic. In DFT, the real electrons of a system are replaced by an effective electrons with the same charge, mass, and density distribution. The real electrons are correlated with other electrons of the system. In contrast to this, the effective electron, as considered in DFT, moves as independent particles in the effective periodic potential [16]. This sets the ground to exploit Bloch's theorem in studying the crystalline solids. The theorem is defined as [3]:

Bloch's Theorem: The eigenstates ψ of the one-electron Hamiltonian H=-$\hbar^2\nabla^2/2m$+$U(\mathbf{r})$, where $U(\mathbf{r}$+$\mathbf{R})$=$U(\mathbf{r})$ for all **R** in the Bravais lattice, can be chosen to have the form of a plane wave times a function with the periodicity of the Bravais lattice [3]:

$$\psi_{n\mathbf{k}}(\mathbf{r}) = e^{i\mathbf{k}.\mathbf{r}} u_{n\mathbf{k}}(\mathbf{r}) \tag{3.4}$$

where $u_{n\mathbf{k}}(\mathbf{r} + \mathbf{R})$=$u_{n\mathbf{k}}(\mathbf{r})$ for all **R** in the Bravais lattice.

Based on the above discussion, it is clear that DFT requires to represent the Bloch states in an appropriate basis set. As we already know that the states corresponding to free electrons are plane wave. Hence, the plane waves basis set can be an intuitively suggested basis set for solving the Kohn-Sham equation. In this direction, it is also seen that Bloch's states are $u_{n\mathbf{k}}(\mathbf{r})$ times $e^{i\mathbf{k}.\mathbf{r}}$. As it involves plane wave, the plane waves can be used as basis set for solving the Hamiltonian. However, in the DFT picture of solids, there exists a drawback of using the plane waves as basis set. This is discussed below.

Drawback: Let us reanalyze the picture of solid as positive ions sitting at their respective sites and valence electrons distributed all over the space under mean-field approximation. For instance, take the case of Calcium (Ca) atom. Its valence electron is $3s$. Now, however the portion of radial part of $3s$ wavefunction that is away from nucleus show smooth variation, it exhibits steep variation in the regions close to nucleus. So, for describing this sharp/steep varying part, a very large number of plane waves will be required. This will not be computationally effective approach. Thus, one cannot rely on plane waves in exploring the regions closer to the nucleus of atoms in a given material using DFT. The most sharp regions of the wavefunction is its tail, situated very close to nucleus. Moreover, this portion of solid is more or less shielded from the outer regions of atoms where chemistry happens. Thus, this portion of atoms will negligibly affect most of the transport and other related properties of solid. Hence, the potential in the regions very close to nucleus (where tail of the wavefunction exists) can be approximated by a smoothly varying pseudopotential function. This will reduce the number of plane waves required for expanding the wavefunction in the sharp varying regions closer to nucleus of atoms in solid. This is exactly what is done in pseudopotential method based DFT packages such as Abinit and Quantum Expresso.

Before concluding this section, we want to highlight that for computational efficiency, the basis set must have following properties [4]:

- *Efficient*: The basis set must be as small in size as possible.
- *Unbiased*: It must not in a hidden way force the solution in a particular direction. In that case, for any new system, it will not be efficient.
- It must be *mathematically simpler*.

3.4 Pseudopotential

As discussed in the previous section, near the nucleus, the orbitals are very localized and hence more number of plane waves are required to find the eigenstates there. However, most of the chemistry in which one is usually interested occurs in the valence region. Thus, for most properties, the electrons in the core regions are almost screened. It is important to note here that when we say core regions in the discussion of *pseudopotential*, we do not explicitly mean core electrons. Instead, we mean the regions close to the nucleus of atoms where even the wavefunction of valence states (like 3*s*) are highly oscillating. Moving further, as the core regions do not affect most of the transport and other related properties of solids to a considerable extent, it is not necessary to capture the exact picture of these regions. This is achieved by creating a *pseudopotential* which is a smoothly varying potential in the core region. However, in the valence region, the *pseudopotential* exactly matches with true potential. In this way, the region around an atom in solid is divided into two parts [4]: (i) the core region where smooth potential is generated and (ii) the valence region where the true potential is taken into account. Introduction of *pseudopotential* ensures that now, the number of plane waves required in solving the Kohn-Sham equation in the core regions will be extremely reduced. This will assure the computational efficiency. It is important to highlight here that situation becomes much more complicated when the material contains 3*d* or 4*f* as valence electrons. This is because 3*d* or 4*f* electrons are spatially more localized. However, being a valence electrons, they may have active participation in bonding. Thus, depending on whether they actively participate in bonding or not, they may be smoothened or not. If smoothening is not possible, then designing an appropriate pseudopotential becomes a challenging task. Also, in the later case, the required number of plane waves increases and thus, the process becomes computationally costly.

There are infinite possible ways of building the *pseudopotentials*. The only requirements are that: (i) it should be *soft*: require very less number of plane waves in the core regions, and (ii) it must be *transferable*: can be used corresponding to a given atom under any circumstances of the environment in which it placed. An efficient *pseudopotential* is one which is both *soft* and *transferable* at the same time.

Drawback: This method will fail to capture the correct picture of the effects that are associated with the states in the core regions or even with the core states [4]. For instance, the hyperfine fields or core level excitation. In such situation, the use of *pseudopotential* to reduce the required number of plane waves is not justified. This demands for the introduction of new types of basis set which is efficient and at the same time describe the core picture with acceptable accuracy.

3.5 Augmented Plane Waves (APW)

As already highlighted, the electrons closer to the nucleus behave more or less like in the free atom. Thus, they can be represented in terms of atomic like functions in a more efficient way. This approach is used to design Augmented Plane Waves (APW) basis set [13]. The APW method was developed by John C. Slater, which later evolved into the more commonly used Linearized APW (LAPW) method (will be discussed in later section). In APW method, around each atom with index α, space is divided into two regions: (i) spherical region surrounding the atoms and containing it, known as *muffin tin region* S_α and (ii) voids between the neighboring *muffin tins*, known as *interstitial regions* denoted as I. It is important to highlight here that such division of space around each atoms is also followed in the Linear Muffin- Tin Orbital (LMTO) method will be discussed in later section. The spherical *muffin tins* are generally kept as large as possible, such that they are non-overlapping but just touching with the *muffin-tin* sphere of the neighboring atom of the solid. However, this is not a strict requirement and thus, one may also change the radius of *muffin tins* if needed. One augmented plane wave (APW) used in the expansion of $\psi_{n\mathbf{k}}(\mathbf{r})$ around an atom with index α is defined as [4]:

$$\phi_{\mathbf{K}}^{\mathbf{k}}(\mathbf{r}, E) = \begin{cases} \frac{1}{\sqrt{V}} e^{i(\mathbf{k}+\mathbf{K})\cdot\mathbf{r}} & \mathbf{r} \in I \\ \sum_{\ell,m} A_{\ell m}^{\alpha,\mathbf{k}+\mathbf{K}} u_\ell^\alpha(r', E) Y_m^\ell(\hat{r}') & \mathbf{r} \in S_\alpha. \end{cases} \tag{3.5}$$

The symbols $\mathbf{K}$ and V stand for reciprocal lattice vector and volume of unit cell respectively. Other symbols have their usual meaning. The position inside the spheres is given with respect to the center of each sphere by $\mathbf{r}' = \mathbf{r} - \mathbf{r}_\alpha$. The length of $\mathbf{r}'$

is r´, and the angles θ' and ϕ' specifying the direction of **r**´ in spherical coordinates are indicated as $\hat{r}'$. The Y_m^ℓ are spherical harmonics. The $A_{\ell m}^{\alpha,\mathbf{k}+\mathbf{K}}$, and E are unknown parameters and need to be determined. Here, E has dimension of energy. The $u_\ell^\alpha(r', E)$ are the solution of radial part of Schrodinger equation for a free atom α at energy E. The term $A_{\ell m}^{\alpha,\mathbf{k}+\mathbf{K}}$ is determined by using the condition that at the boundary of the muffin tin sphere, the value of function in the sphere should match with the plane wave at each point of the sphere. Thus, one can see APW as some oscillating function which turns into more complex form when it comes closer to the nucleus of an atom. So far, it seems that APW can serve the purpose of solving the Kohn-Sham equation with needed efficiency. However, it is not. Now, we shall discuss its drawback.

Drawback: For representing $\psi_{n\mathbf{k}}(\mathbf{r})$ with APW, one needs to set E parameter equal to its eigenenergy $\varepsilon_{n\mathbf{k}}$. However, this is exactly what is needed to be determined. Hence, the problem becomes of a kind which requires E to be self-consistently determined for each $\psi_{n\mathbf{k}}(\mathbf{r})$ [4]. Thus, with the planewaves, N eigenvalues are found by a single diagonalization (and following self-consistent method). However, with APW, each eigenvalue demands separate self-consistent diagonalization. This makes the method much slower than the *pseudopotential* method.

3.6 Linearized Augmented Plane Wave (LAPW)

3.6.1 The Regular LAPW Method

Linearized Augmented Plane Wave (LAPW) basis set was first proposed by Ole Krogh Andersen. Here, we shall discuss LAPWs in details. Considering the drawbacks of APW approach, it will be better if one could find $u_\ell^\alpha(r', E)$ from some quantity which is known beforehand. Consider that value of $u_\ell^\alpha(r', E)$ is known at E_0, one can find its value at neighboring values of E through Taylor series expansion as shown below [4]

$$u_\ell^\alpha(r', \epsilon_{\mathbf{k}}^n) = u_\ell^\alpha(r', E_0) + \underbrace{\left(E_0 - \epsilon_{\mathbf{k}}^n\right) \left.\frac{\partial u_\ell^\alpha(r', E)}{\partial E}\right|_{E=E_0}}_{i_\ell^\alpha(r', E_0)} + O\left(E_0 - \epsilon_{\mathbf{k}}^n\right)^2. \tag{3.6}$$

The substitution of first two terms of above expansion in the APW basis set gets us close to what is known as LAPW basis set. In this process, the value of $O\left(E_0 - \epsilon_{\mathbf{k}}^n\right)$ is unknown leading to the addition of another undetermined coefficient $B_{\ell m}^{\alpha,\mathbf{k}+\mathbf{K}}$. Thus, LAPW basis set is preliminary defined as [4]-

$$\phi_{\mathbf{K}}^{\mathbf{k}}(\mathbf{r}) = \begin{cases} \frac{1}{\sqrt{V}} e^{i(\mathbf{k}+\mathbf{K})\cdot\mathbf{r}} & \mathbf{r} \in I \\ \sum_{\ell,m} \left(A_{\ell m}^{\alpha,\mathbf{k}+\mathbf{K}} u_{\ell}^{\alpha}(r', E_0) + B_{\ell m}^{\alpha,\mathbf{k}+\mathbf{K}} \dot{u}_{\ell}^{\alpha}(r', E_0) \right) Y_m^{\ell}(\hat{r}') & \mathbf{r} \in S_{\alpha}. \end{cases} \tag{3.7}$$

The coefficients $A_{\ell m}^{\alpha,\mathbf{k}+\mathbf{K}}$ and $B_{\ell m}^{\alpha,\mathbf{k}+\mathbf{K}}$ are obtained by requiring the condition that function inside the sphere should match to the plane wave at boundary in terms of both value and the slope.

Nextly, suppose one is trying to represent an eigenstate $\psi_{n\mathbf{k}}(\mathbf{r})$ which is majorly contributed by s-character ($l = 0$). Then, in its expansion in LAPW, the term $A_{(\ell=0)m)}^{\alpha,\mathbf{k}+\mathbf{K}}$ will be large. In that case, it will be an efficient approach to set E_0 at the energy corresponding to center of s-band. In this way, the term $O\left(E_0 - \epsilon_{\mathbf{k}}^n\right)$ will be relatively very small and its exclusion will be allowed. This process can be repeated to determine E corresponding to each l ($l = 0, 1, 2$, and 3) and for all atoms. For higher value of l, a fixed E can be assigned. Thus, the final definition of LAPW basis set is [4]

$$\phi_{\mathbf{K}}^{\mathbf{k}}(\mathbf{r}) = \begin{cases} \frac{1}{\sqrt{V}} e^{i(\mathbf{k}+\mathbf{K})\cdot\mathbf{r}} & \mathbf{r} \in I \\ \sum_{\ell,m} \left(A_{\ell m}^{\alpha,\mathbf{k}+\mathbf{K}} u_{\ell}^{\alpha}(r', E_{0,l}^{\alpha}) + B_{\ell m}^{\alpha,\mathbf{k}+\mathbf{K}} \dot{u}_{\ell}^{\alpha}(r', E_{0,l}^{\alpha}) \right) Y_m^{\ell}(\hat{r}') & \mathbf{r} \in S_{\alpha}. \end{cases} \tag{3.8}$$

Once the value of E_{0l} is fixed (it is generally taken as 0.30 Rydberg in WIEN2k [4]), the basis functions can be calculated for all the states $\psi_{n\mathbf{k}}(\mathbf{r})$ at a given $\boldsymbol{k}$ in one shot. Thus, now the case becomes similar to plane waves in terms of the fact that in a single diagonalization, N different band energies can be determined at a given $\boldsymbol{k}$.

Drawback: Unlike plane wave basis functions, which are orthogonal by design, LAPW basis functions are not orthogonal [4]. As a result, the overlap matrix (S) must be explicitly calculated and included in the eigenvalue problem:

$$H\Psi = ES\Psi$$

where H is the Hamiltonian, S is the overlap matrix, and Ψ is the wavefunction. The need to compute and store S increases memory and computational requirements. Moreover, the non-orthogonality transforms the standard eigenvalue problem into a *generalized eigenvalue problem*, which is more computationally intensive to solve:

$$H\Psi = ES\Psi.$$

Solving the generalized eigenvalue problem scales poorly with the size of the basis set compared to the orthogonal case.

3.6.2 LAPW with the Local Orbitals (LAPW+LO)

In this method [4], it is emphasized that the states which are very close to the nucleus, for instance $1s$ in Fe, are considered as core state. Such states are entirely contained within the muffin tin and thus stay away from chemical bonding with the neighboring atoms of solid. These states will behave exactly as if they were in the free Fe atom. Thus, they are not required to be treated with LAPW. The states leak-out of the muffin-tin participate in bonding and are termed as valence state. These states are required to be treated using LAPW basis set. Core states are treated as the states in free atoms under the effect of potential due to the valence states.

In the LAPW method, there is a possibility of the existence of two states, for instance $3p$ and $4p$ in Fe, such that both of them leak-out of muffin-tin and hence having considerable contribution in the valence states. This makes it unclear as to what value must be assigned to $E^{Fe}_{0(l=1)}$: close to $3p$ or close to $4p$ or some other value. This problem is solved by adding *local orbitals* (LO) basis functions to LAPW. These *LOs* are defined as [4]-

$$\phi^{lm}_{\alpha,LO}(\mathbf{r}) = \begin{cases} 0 & \mathbf{r} \notin S_\alpha \\ \left(A^{\alpha,LO}_{\ell m} u^\alpha_\ell(r', E^\alpha_{1,l}) + B^{\alpha,LO}_{\ell m} \dot{u}^\alpha_\ell(r', E^\alpha_{1,l}) + C^{\alpha,LO}_{\ell m} u^\alpha_\ell(r', E^\alpha_{2,l})\right) Y^\ell_m(\hat{r}') & \mathbf{r} \in S_\alpha. \end{cases} \tag{3.9}$$

Properties of Local Orbital [4]:

- They are defined for particular l and m and for a given atom α.
- Their value is zero in the interstitial regions and also in the muffin-tin region of other atoms. Hence, termed as *local orbitals*.
- They are not connected to the plane waves in the interstitial regions. Thus, they are not dependent on $\mathbf{k}$ or $\mathbf{K}$ vectors.

The coefficients $A^{\alpha,LO}_{\ell m}$, $B^{\alpha,LO}_{\ell m}$, and $C^{\alpha,LO}_{\ell m}$ are obtained by using the conditions that LO is normalized, and has zero value and zero slope at the boundary of the sphere [4]. Adding LO slightly increases the basis size of LAPW basis set. However, it enhances the numerical accuracy of the results. The LAPW+LO method is one of the most accurate all-electron methods in DFT, often referred to as the "gold standard" for solid-state calculations. Hence, it is very demanding in the present DFT community.

3.6.3 *Muffin-Tin Orbital (MTO) and Linear Muffin-Tin Orbital (LMTO) Basis in Terms of Bessel and Hankel Functions*

The idea of Muffin-Tin Orbital (MTO) was put forward by Ole Krogh Andersen in 1971 [5]. The MTO basis set comprises of localized augmented orbitals. As in the case of local orbitals, the Bloch states are supposed to be expressed in small number of meaningful functions (MTOs). However, unlike local orbitals, the MTOs are more accurate as they are generated from Kohn-Sham Hamiltonian itself. These basis sets are most efficient for closed packed solids.

3.6.3.1 Muffin-Tin Orbital (MTO) Basis

The Muffin-Tin Orbital (MTO) basis function is designed to describe solutions of the Schrödinger equation inside the muffin-tin spheres and in the interstitial region. These are localized basis functions continuous in value and derivative at the muffin-tin sphere boundary [6].

1. **Inside the Muffin-Tin Sphere**: Inside the muffin-tin sphere of radius R_{MT}, the wave function (solution of the Kohn-Sham Hamiltonian) is expressed in terms of **spherical Bessel functions** $j_l(kr)$, which are regular at $r = 0$:

$$\phi^{\text{MT}}_{\mathbf{k}+\mathbf{G}}(\mathbf{r}) = \sum_{l,m} A_{lm} u_l(r) Y_{lm}(\hat{\mathbf{r}}),$$

   where

   - $u_l(r)$ is the radial solution of the Schrödinger equation or its approximation.
   - $Y_{lm}(\hat{\mathbf{r}})$ is spherical harmonics describing the angular dependence.
   - $u_l(r)$ is expanded using the spherical Bessel function:

$$u_l(r) = j_l(kr),$$

   where k is the wavevector magnitude inside the sphere. Note, that there is no summation over k because here, k is the magnitude of the wavevector corresponding to the energy of the electron state being considered. l is the angular momentum quantum number. This direct relationship comes from the fact that $j_l(kr)$ is an exact solution to the radial Schrödinger equation for a spherical potential when the energy (and hence k) is fixed. In other words, inside the muffin-tin sphere, the wavefunction assumes the spherical symmetry, and for a specific k, the radial solution is uniquely determined by the spherical Bessel function $j_l(kr)$.

2. **In the Interstitial Region**: In the interstitial region, where the potential is approximately constant, the wave function is described by a **spherical Hankel function** $h_l^{(2)}(kr)$, which represents outgoing spherical waves:

$$\phi_{\mathbf{k}+\mathbf{G}}^{\text{interstitial}}(\mathbf{r}) = h_l^{(2)}(kr)Y_{lm}(\hat{\mathbf{r}}).$$

3.6.3.2 Linear Muffin-Tin Orbital (LMTO) Basis

The LMTO basis is a linearized version of the MTO basis, designed for computational efficiency and simplification.

- **Linearization in Energy**: The radial part of the wave function $u_l(r, E)$ is linearized around a fixed reference energy E_ν:

$$u_l(r, E) \approx u_l(r, E_\nu) + (E - E_\nu)\dot{u}_l(r, E_\nu),$$

  where

  - $u_l(r, E_\nu)$ is the radial solution at the reference energy.
  - $\dot{u}_l(r, E_\nu)$ is the energy derivative of the radial solution.

- **Inside the Muffin-Tin Sphere**: The radial solution $u_l(r)$ is represented as a **spherical Bessel function** $j_l(kr)$ and its energy derivative:

$$u_l(r, E) \approx j_l(kr) + (E - E_\nu)\frac{\partial j_l(kr)}{\partial E}.$$

- **In the Interstitial Region**: In the interstitial region, the solution is matched to a **spherical Hankel function** $h_l^{(2)}(kr)$, representing outgoing waves:

$$\phi_{\text{LMTO}}^{\text{interstitial}}(\mathbf{r}) = h_l^{(2)}(kr)Y_{lm}(\hat{\mathbf{r}}).$$

- **Combined LMTO Basis**: The LMTO basis combines the radial and angular components inside the muffin-tin sphere with the asymptotic behavior in the interstitial region:

$$\phi_{\mathbf{k},l,m}(\mathbf{r}) = \begin{cases} j_l(kr)Y_{lm}(\hat{\mathbf{r}}) + (E - E_\nu)\frac{\partial j_l(kr)}{\partial E}Y_{lm}(\hat{\mathbf{r}}), & r \le R_{\text{MT}}, \\ h_l^{(2)}(kr)Y_{lm}(\hat{\mathbf{r}}), & r > R_{\text{MT}}. \end{cases}$$

3.6.3.3 Physical Meaning of Bessel and Hankel Functions

- **Bessel Functions** $j_l(kr)$:
  - solve the Helmholtz equation inside the muffin-tin sphere,
  - represent **standing waves** that are finite at the origin,
  - describe the radial part of the wavefunction confined within the spherical region around atoms.
- **Hankel Functions** $h_l^{(2)}(kr)$:
  - represent **outgoing spherical waves** in the interstitial region (outside the muffin-tin sphere),
  - used to ensure continuity of the wavefunction and its derivative across the muffin-tin boundary.

Summary: These functions provide a *compact representation of wave functions* by exploiting the spherical symmetry of the muffin-tin approximation in crystalline solids, simplifying the solution of the Schrödinger equation.

For more detailed discussion regarding MTO/LMTO methods, one should refer the book written by Richard M. Martin [7].

3.7 Full-Potential Method

In previous discussion, we have seen that APW and LAPW methods expand the wave function inside the muffin-tin in the form of spherical harmonics multiplied by $u_\ell^\alpha(r', E)$ (and its derivatives). In this approach, the $u_\ell^\alpha(r', E)$ preassumes the potential inside the atomic spheres (muffin-tin spheres) to be spherically symmetric. This approximation simplifies the calculations but limits the accuracy when dealing with systems where the potential deviates significantly from spherical symmetry. In this context, the "*full potential*" method removes the spherical symmetry approximation, thus making the method more accurate [1]. Here, we shall discuss two full potential methods: (I) the *full-potential LAPW (FP-LAPW)* method and (II) the *APW+local orbital (lo)* method.

3.7.1 Full-Potential LAPW (FP-LAPW) [8]

The FP-LAPW method builds upon LAPW by allowing for the potential to be non-spherical inside the muffin-tin spheres. This introduces additional flexibility in the basis. The FP-LAPW basis functions are defined as-

Table 3.1 Comparison of features of LAPW and FP-LAPW methods

Feature	LAPW	FP-LAPW
Potential treatment	Spherically symmetric	Fully non-spherical
Basis function flexibility	Limited	Higher
Accuracy	Moderate	High
Computational cost	Lower	Higher
Applications	Simple bulk systems	Complex systems (surfaces, anisotropic materials)

$$\phi_{\mathbf{K}}^{\mathbf{k}}(\mathbf{r}) = \begin{cases} \frac{1}{\sqrt{V}} e^{i(\mathbf{k}+\mathbf{K})\cdot\mathbf{r}} & \mathbf{r} \in I \\ \sum_{\ell,m} \left(A_{\ell m}^{\alpha,\mathbf{k}+\mathbf{K}} u_{\ell,m}^{\alpha}(r', E_{0,l}^{\alpha}) + B_{\ell m}^{\alpha,\mathbf{k}+\mathbf{K}} \dot{u}_{\ell,m}^{\alpha}(r', E_{0,l}^{\alpha}) \right) Y_m^{\ell}(\hat{r}') & \mathbf{r} \in S_{\alpha}. \end{cases} \tag{3.10}$$

It is seen from Eq. 3.10 that now the $u_{\ell,m}^{\alpha}(r', E_{i,l}^{\alpha})$ depends on both r and m. This allows the consideration of non-spherical contributions. This refinement significantly improves the accuracy of the representation of wave functions near the nucleus. Thus, FP-LAPW is more efficient and suggested method for exploring complex systems (surfaces, anisotropic materials) over LAPW. It must be noted here that, outside the muffin-tin sphere, the basis set in FP-LAPW is same plane waves as used in the LAPW method. A comparison of features of LAPW and FP-LAPW methods is presented in Table 3.1.

3.7.2 APW+lo Basis Set

3.7.2.1 Pure APW+lo Basis Set

The APW+lo basis set comprises of two kinds of functions. The first kind are APWs with set of fixed energy parameter $E_{1,l}^{\alpha}$. This is given as [4]

$$\phi_{\mathbf{K}}^{\mathbf{k}}(\mathbf{r}, E) = \begin{cases} \frac{1}{\sqrt{V}} e^{i(\mathbf{k}+\mathbf{K})\cdot\mathbf{r}} & \mathbf{r} \in I \\ \sum_{\ell,m} A_{\ell m}^{\alpha,\mathbf{k}+\mathbf{K}} u_{\ell}^{\alpha}(r', E_{1,l}^{\alpha}) Y_m^{\ell}(\hat{r}') & \mathbf{r} \in S_{\alpha}. \end{cases} \tag{3.11}$$

This method involves full-potential calculation via recalculating u_{ℓ}^{α} in each SCF cycle, allowing for an expansion/contraction corresponding to the given charge state (ionicity) of the atom [12]. Thus, it does not preassumes the shape of the potential inside the muffin-tin sphere. Furthermore, to make it efficiently work with fixed value of $E_{1,l}^{\alpha}$, it is combined with another set of *local orbitals*. It is important to note that $E_{1,l}^{\alpha}$ are precalculated values of E parameters for different values of l for various atoms. This can be used as such in calculations. Moreover, the local orbitals used in this method are different from the one used in LAPW+LO in functional form. Hence,

Table 3.2 Comparison of features of FP-LAPW and APW+lo methods

Feature	FP-LAPW	APW+lo
Basis set	Augmented plane waves	Augmented plane waves + local orbitals (los)
Tightly bound states	Requires high cutoff	Handled efficiently with los
Computational cost	Higher	Moderate
Accuracy	High	High (with reduced size of basis set)
Typical applications	General-purpose	Systems with semicore states

these *local orbitals* are abbreviated as *lo* and are defined as [4]

$$\phi^{lm}_{\alpha,lo}(\mathbf{r}) = \begin{cases} 0 & \mathbf{r} \notin S_\alpha \\ \left(A^{\alpha,lo}_{\ell m} u^\alpha_\ell(r', E^\alpha_{1,l}) + B^{\alpha,lo}_{\ell m} \dot{u}^\alpha_\ell(r', E^\alpha_{1,l})\right) Y^\ell_m(\hat{r}') & \mathbf{r} \in S_\alpha. \end{cases} \tag{3.12}$$

Here, $E_{1,l}$ can be same set of energies as is used in defining APW. However, other set of energies can also be used [4]. The coefficients $A^{\alpha,lo}_{\ell m}$ and $B^{\alpha,lo}_{\ell m}$ can be obtained by normalization condition and by using the fact that local orbital has zero value at the boundary of the muffin tin. It is important to note that for a given level of accuracy, APW+lo require almost same number of basis set as for the case of APW. This number is smaller when compared with the LAPW+LO method and hence APW+lo is computationally more efficient. In addition to this, N number of eigenvalues can be obtained in a single diagonalization using the APW+lo method as in LAPW+LO. A comparison of features of FP-LAPW and APW+lo methods is presented in Table 3.2.

3.7.2.2 Mixed LAPW/APW+lo Basis Sets

LAPW cannot efficiently treat two cases: (I) Valence *d*- and *f*-states and (II) states in atoms that have muffin-tin sphere comparatively much smaller than sphere associated with other atoms in the unit cell [4]. Thus, most efficient approach will be to treat these states using APW+lo orbitals while the other states with the LAPW. Such a mixed basis set is found to be highly efficient and thus recommended choice of DFT packages like WIEN2k and Elk.

The Merging Distinction Between LAPW and APW+lo Methods [17]

The distinction between LAPW and APW+lo has indeed become increasingly blurred, primarily due to advancements in computational methods and improvements in algorithms implemented in modern electronic structure codes like WIEN2k and Elk. The LAPW method traditionally uses a linearized approach to describe wavefunctions in the muffin-tin spheres. As previously discussed, modern LAPW implementations often include LOs to better describe semicore states. This inclusion of local orbitals makes LAPW increasingly similar to APW+lo, which explicitly combines the APW basis with additional local orbitals for flexibility and accuracy. Nextly, APW+lo allows for the explicit dependence of basis functions on energy, avoiding linearization errors inherent to traditional LAPW. However, modern LAPW implementations also mitigate these errors by incorporating multiple local orbitals and optimizing the basis set for better variational convergence, reducing the practical differences. Codes like WIEN2k and Elk effectively unify the LAPW and APW+lo methods, offering users the flexibility to combine features of both approaches efficiently. Hence, in present times, the distinctions are more in terms of how basis sets are constructed rather than in the theoretical or practical outcomes.

3.8 Convergence Criteria

The Kohn-Sham equation is solved self-consistently to obtain converged ground state density. The accuracy of the calculated properties depends on the value of various parameters used while solving the Kohn-Sham equation self-consistently. These parameters are discussed below

1. **Energy Cutoff (Ecut)**: This controls the number of plane waves that must be used in solving the Kohn-Sham equation. The allowed plane waves are those with kinetic energy lower than ecut, which translates to the following constraint on the planewave vector $\boldsymbol{G}$ in reciprocal space is decided by [18]

$$\frac{1}{2}(2\pi)^2(\boldsymbol{k}+\boldsymbol{G})^2 < ecut. \tag{3.13}$$

   Higher value of ecut will increase the number of plane waves. More number of plane waves will ensure to capture the finer details of the wave function and hence the solid. However, this will in turn increase the computational cost. Thus, it is needed to converge the *Ecut* value such that change in the property under study with further increase in *Ecut* is negligible to the required interest. This approach is only followed in the DFT packages which uses plane waves as basis set (for instance, Abinit, Quantum Expresso, etc.). The APW+lo or full potential packages like WIEN2k and Elk cannot use this approach. The idea followed in full potential packages is discussed below.

For the APW or LAPW basis set, it is better to judge the accuracy of result based on the product $R_{min}G_{max}$ between the smallest muffin tin radius and G_{max} [4]. This process is explicitly followed in WIEN2k package. Let us explain it briefly. If the smallest muffin tin radius is increased, the closest point a plane wave can come to a nucleus moves farther away from the nucleus. The part of the wave function that need not to be described with plane waves any more in general will have displayed the steepest behavior, steeper than anywhere else in the interstitial region (it was closest to the nucleus). Less plane waves are needed to describe the remaining, smoother parts of the wave function. Thus, G_{max} can be reduced. In this method, the DFT community generally follow that the product $R_{min}G_{max}$ should remain constant in order to have comparable accuracy. Reducing G_{max} means reducing the size of the matrices, and because matrix diagonalization is very expensive, a larger R_{min} can significantly reduce the computation time. R_{min} cannot be too large on the other hand, as the spherical harmonics are not suited to describe the wave functions in the region far away from the nuclei. One must note here that the Elk package (which also uses APW basis set) uses a slightly different criterion. While defining the product mentioned above, it uses R as mean value of all the muffin-tin radii. The other explanation remain the same.

2. **$\boldsymbol{k}$-point Sampling**: The $\boldsymbol{k}$-point sampling is preferred to be Γ-centered. This is found to generally reduce the computational cost and contains the important information regarding the band gaps. The increase in the number of $\boldsymbol{k}$-points sampling enhances the accuracy of the result. Thus, convergence of the property under study with respect to $\boldsymbol{k}$-point is also needed. It is important to note that various property requires different levels of the density of $\boldsymbol{k}$-mesh gridding across the Brillouin zone (BZ). For instance, a converged ground state energy to an acceptable accuracy is generally obtained with about $10\times10\times10$ $\boldsymbol{k}$-mesh while the transport properties like electrical conductivity may require $50\times50\times50$ or even more denser $\boldsymbol{k}$-mesh gridding across the BZ. Thus, any property under study must be separately ensured to be converged with respect to the number of $\boldsymbol{k}$-points sampled across the full BZ.
3. **SCF Convergence Criteria**: Having ensured the above two convergence criteria, the convergence must be tested in the total energy and the charge density obtained in the two consecutive SCF cycle. Generally, the difference in the energy and charge density between two consecutive SCF cycle must be less than 10^{-6} Rydberg and 10^{-4} electrons/bohr^3, respectively. However, these criteria also depend on the properties under study. For instance, for phonon calculations, the aforementioned criteria may not be sufficient and more strict criteria may be required.
4. **Force Convergence Criteria**: This ensures the accurate atomic positions and cell geometries of the crystal under study. Hence, the convergence is very much needed.
5. **Supercell Size**: For defect or surface calculations, the supercell must be large enough to minimize interactions between periodic images.

6. **Number of Unoccupied Bands**: In DFT, the number of unoccupied bands is crucial for accurate results in calculations involving electronic excitations (e.g., optical properties, Berry curvature, GW), as these bands capture transitions and higher-energy states. Insufficient bands can lead to errors in total energy, dielectric response, and quasiparticle energies. Always ensure convergence by including enough unoccupied states, especially for properties requiring interband transitions or many-body effects.
7. **Vacuum Thickness**: In DFT, the 2D calculations or surface states calculation is generally performed by forming a supercell by growing the unit cell in one crystal direction (let's say in the c direction). Then, the addition of vacuum is needed among the neighboring unit cells in this direction. This is done to reduce the interaction between the layers of these unit cells. To ensure minimal interaction between these layers, convergence of results (line surface states) must be carried out with respect to vacuum thickness.

3.9 Practical Tips for Setting-Up DFT Calculations

Firstly, we shall discuss the pre-calculation setup which is a crucial step in any DFT simulation. Properly preparing the system and choosing appropriate computational parameters can save significant time and resources while ensuring reliable and accurate results. This section covers essential aspects of setting up a DFT calculation, including structural optimization, choice of exchange-correlation functional, and initial parameter selection.

3.9.1 Exchange-Correlation Functional [7]

The exchange-correlation (XC) functional is a central component of DFT, as it helps in correcting the discrepancy created in correct picture of electron-electron interaction on moving from many-particle Hamiltonian to a Kohn-Sham Hamiltonian representing single electron in a periodic potential. Selecting the appropriate XC functional is critical for obtaining accurate results.

Common XC Functionals

- **Local Density Approximation (LDA)**: Accurate for studying bulk materials with relatively uniform electron density, like metals or simple semiconductors.
- **Generalized Gradient Approximation (GGA)**: Includes gradient corrections, improving accuracy for systems with varying electron densities. A commonly used functional is Perdew-Burke-Ernzerhof (PBE) [14].

- **Hybrid Functionals** (e.g., HSE06, PBE0): Incorporate a portion of exact exchange from Hartree-Fock theory, providing improved accuracy for band gaps and excited states, though they are computationally expensive.
- **The Tran-Blaha modified Becke-Johnson (TB-mBJ)**: It is very useful in case where one needs to accurately predict the band gap of a material, especially for semiconductors and insulators. It tends to provide more accurate band gap values compared to standard functionals like GGA, which often underestimate the gap.
- **Meta-GGA**: Meta-GGAs often provide better predictions of band gaps compared to standard GGAs. In addition to this, it is very useful for studying systems involving a mixture of different types of chemical bonds (e.g., metallic and covalent).

Selecting the Right XC Functional

- **For Metals**: Use LDA or GGA, as hybrid functionals may overestimate metallic properties.
- **For Semiconductors and Insulators**: Use hybrid functionals to capture accurate band gaps. Apart from this, TB-mBJ is also very effective in addressing the band gap and hence can be a preferred choice.
- **For Strongly Correlated Systems**: Hybrid functional is considered to be good. However, the DFT+U or dynamical mean-field theory (DMFT) methods are generally preferred.

Validation

- Compare calculated properties (e.g., lattice constants, band gaps, etc.) with experimental data or higher-level theoretical methods.

3.9.2 Structural Relaxation

Structural relaxation ensures that the atomic positions in a system correspond to the minimum energy configuration. This step is especially important for materials where experimental atomic coordinates are not available or where the initial structure might not be optimal.

Why Structural Relaxation Is Important

- Forces on atoms and stresses in the lattice can result in unstable or unrealistic configurations if unrelaxed structures are used.

- Relaxation eliminates non-physical strains and ensures that electronic properties are calculated for the most stable geometry.
- This is especially important for phonon calculations and the related properties.

Types of Relaxation

- **Atomic Relaxation**: Adjusts atomic positions while keeping the cell parameters fixed. Suitable for molecules and surfaces.
- **Cell Relaxation**: Adjusts both atomic positions and lattice parameters. Necessary for bulk materials and systems under stress.

Convergence Criteria for Relaxation

- Typical thresholds for atomic forces are < 0.01 eV/Å.

Practical Considerations

- Use a moderate k-point grid and plane-wave cutoff during initial relaxation steps. Refine these parameters for the final electronic structure calculation.
- Ensure that symmetry constraints are appropriately applied or lifted depending on the nature of the system (e.g., defects or amorphous structures may require breaking symmetry).

3.9.3 Initial Parameter Selection

Selecting appropriate computational parameters at the outset can significantly influence the efficiency and accuracy of DFT calculations.

Use appropriate size of basis set.

- For plane wave based packages: plane-wave cutoff energy (E_{cut})
  - Ensure the cutoff energy is sufficient to represent the wave functions accurately.
  - Start with a value suggested by the pseudopotential and refine through convergence testing.
- For APW or LAPW based packages: One must set appropriate value of $R_{min}G_{max}$. For packages like WIEN2k, values between 5.0–9.0 (3.0 if there are small H-spheres) are usually reasonable[12].

k-Point Grid Density

- Define a Monkhorst-Pack grid based on the system's periodicity:
  - For the ground state self-consistent density calculations, a dense grid is needed for metals (e.g., $12 \times 12 \times 12$) whereas one can go with the coarser grids for insulators (e.g., $6 \times 6 \times 6$).
  - For the calculations of transport properties, generally a very dense k-mesh is needed. This may vary from ($50 \times 50 \times 50$) to ($400 \times 400 \times 400$) or even higher depending on the package used and the material under study.
  - For supercell calculations (e.g., surface states or phonons using the supercell approach), one generally requires smaller k-mesh size. This is because as the size of supercell is higher than the primitive cell or unit cell, its corresponding BZ will be smaller. Thus, smaller size of k-grid will be enough to capture details to a required accuracy.
- In all DFT calculations, one must always perform the convergence of each property with respect to the size of k-mesh.

3.9.3.1 Charge Mixing Schemes in DFT

Charge mixing schemes are essential techniques in DFT to accelerate the convergence of self-consistent field calculations. These schemes update the electron density (or potential) iteratively to minimize the total energy and achieve self-consistency. Below are the main charge mixing schemes used in DFT:

Linear Mixing

- It is the simplest approach, where the new charge density is a linear combination of the input (current) density and the output (calculated) density.
- **Formula** [7]:

$$\rho_{i+1}^{\text{new}} = (1 - \alpha)\rho_i^{\text{old}} + \alpha\rho_i^{\text{out}}$$

  where α is the mixing parameter ($0 < \alpha < 1$), ρ_i^{out} and ρ_i^{old} are the input and the calculated densities at the ith iteration. The ρ_{i+1} is the charge density at (i+1)th iteration.
- One must note here that the output density of one cycle is not as such taken as input density of next cycle (it mixed with the input density of the present cycle). This is generally done to avoid getting density at any iteration that is far from the resemblance of solid under study.
- It is easy to implement. It is useful for the case where there is oscillatory behavior in the energy during the convergence.

Pulay Mixing (a form of Broyden's Method) [9]

- It is a more advanced scheme that uses a history of past charge densities to predict the optimal update, effectively solving a quasi-linear system for better convergence.
- **Approach:**
  - In this approach, residual is calculated as
    $$\Delta\rho = \rho^{\text{out}} - \rho^{\text{in}}.$$
  - $\Delta\rho$ of the previous iterations is used to improve the update (via a least-squares approach). The expression for the updated density is-
    $$\rho_{i+1}^{\text{new}} = \rho_i^{\text{old}} + \alpha_i \Delta\rho_i$$
    where ρ_i^{old} and ρ_{i+1}^{new} are the output density of the previous iteration and the input density of the present iteration, respectively.
- It accelerates the convergence. It efficiently handles metallic systems.
- **Limitations:** Requires storing previous iterations, which can be memory-intensive for large systems.
- **Note:** Broyden's mixing in its general form is the more advanced method. The Broyden's method is the default mixing scheme for DFT packages like VASP (Vienna Ab initio Simulation Package) and Quantum ESPRESSO.

Anderson Mixing [10]

- It is similar to Pulay mixing but uses a different formulation to optimize convergence.
- **Formula:**
  $$\rho^{\text{new}} = \rho^{\text{old}} + \sum_{i=1}^{m} c_i \Delta\rho_i,$$
  where c_i are coefficients obtained by minimizing residual errors.
- The coefficients c_i in Anderson mixing are determined by solving a least-squares minimization problem. The goal is to minimize the norm of the residuals over the last m iterations:
  $$\min_{c_i} \left\| \Delta\rho - \sum_{i=1}^{m} c_i \Delta\rho_i \right\|^2,$$
  where
  - $\Delta\rho$ is the current residual, defined as $\Delta\rho = \rho^{\text{out}} - \rho^{\text{in}}$,
  - $\Delta\rho_i$ are the residuals from the i-th previous iteration, and
  - c_i are the coefficients that weight the contribution of each previous residual.

To find the optimal coefficients c_i, the least-squares problem is typically formulated as solving a linear system:

$$\mathbf{Ac} = \mathbf{b},$$

where

- $\mathbf{A}$ is the matrix of overlaps between previous residuals, defined as $A_{ij} = \langle \Delta\rho_i, \Delta\rho_j \rangle$,
- $\mathbf{b}$ is the vector of overlaps between the current residual and previous residuals, defined as $b_i = \langle \Delta\rho, \Delta\rho_i \rangle$,
- $\mathbf{c}$ is the vector of coefficients c_i.

By solving this linear system, the coefficients c_i are determined, and the new charge density update is computed as

$$\rho^{\text{new}} = \rho^{\text{old}} + \sum_{i=1}^{m} c_i \Delta\rho_i.$$

- It is effective for systems with complex convergence behaviors.
- **Limitations:** Computationally intensive compared to linear mixing.

Kerker Mixing [11]

- It is specially designed for systems with poor screening, such as metals, where the long-wavelength components of the charge density oscillate slowly.
- **Approach:**
  - Enhances short-wavelength components of the charge density.
  - Reduces long-wavelength components using a weighting function in reciprocal space.
- It is useful for metallic and slab systems.
- **Limitations:** Not suitable for all systems; requires tuning parameters.

Hybrid Mixing Schemes

- This method combines different mixing schemes (e.g., linear + Pulay) to bring together their advantages.
- **Application:**
  - Start with linear mixing for initial iterations.
  - Transition to Pulay or Anderson mixing once the system stabilizes.
- It balances simplicity and convergence efficiency.

Practical Tips for Using Charge Mixing: Almost all the DFT packages provide default values of the parameters for their mixing schemes. It is suggested to stick to these default values. However, if no option is available, one can always perform calculations using smaller value of the mixing parameters. By carefully choosing and optimizing a charge mixing scheme, SCF convergence can be significantly enhanced, saving computational time and resources. It is also important to note that out of the various charge mixing schemes discussed above, linear mixing is claimed to be more stable.

Spin Polarization

- For magnetic systems, enable spin-polarized calculations to account for unpaired electrons.

3.9.4 Points to Be Taken Care of

Before starting the calculations:

1. Verify the input structure for correctness and remove any anomalies (e.g., overlapping atoms).
2. Perform a quick test run with loose convergence criteria to identify potential issues.
3. Cross-check all parameters with the system under study and validate against known benchmarks or literature values.

By thoroughly addressing these aspects, the pre-calculation setup ensures that subsequent DFT calculations proceed smoothly and yield accurate, reproducible results.

3.10 Execution and Post-processing

After the pre-calculation setup, the next phases are executing the DFT simulations and analyzing the results. This section discusses the essential steps involved in running DFT calculations, troubleshooting potential issues during execution, and performing post-processing to extract meaningful insights.

3.10.1 Execution of DFT Calculations

The execution phase involves solving the Kohn-Sham equations self-consistently, ensuring convergence of electronic density and total energy.

3.10.1.1 Running the SCF Loop

The SCF loop solves the Kohn-Sham equations iteratively until the electronic density and energy meet predefined convergence criteria:

- Energy difference: $< 10^{-6}$ Rydberg.
- Charge density difference: $< 10^{-4}$ electrons/bohr^3.

Monitor convergence throughout the process. Oscillations or slow convergence often signal issues such as poor mixing parameters or insufficient k-points.

3.10.1.2 Parallelization

Modern DFT codes allow computations to be parallelized to improve efficiency:

- **Over k-points**: Useful for systems with large Brillouin zones.
- **Over bands**: Efficient for systems with many electronic states.
- **Over Fast Fourier Transform (FFT) grids**: Reduces computational time for plane-wave-based codes.

Optimize the number of processors and the level of parallelization based on the system and resources.

3.10.1.3 Monitoring the Calculation

Regularly check log files to track convergence and detect potential issues such as divergence or non-convergence. Use intermediate checkpoints to restart calculations in case of hardware or software interruptions.

3.10.1.4 Job Scheduling

On high-performance computing (HPC) systems, use job schedulers like SLURM or PBS to manage runs. Specify appropriate resource allocations (e.g., number of cores, memory, and runtime limits) to avoid premature termination.

3.10.2 Troubleshooting Common Issues

Efficient troubleshooting ensures that calculations converge reliably and within reasonable computational time.

3.10.2.1 Divergence in SCF Loop

- **Causes**: Poor mixing parameters, insufficient k-point grid density, or inappropriate initial density.
- **Solutions**:
  - Adjust mixing parameters, e.g., reducing the mixing factor to stabilize updates.
  - Refine k-points or plane-wave cutoff.
  - Use a pre-converged density from a similar system as the starting point.

3.10.2.2 Slow Convergence

- **Causes**: Complex electronic structures (e.g., systems with a near-metallic state) or low smearing width in metallic systems.
- **Solutions**:
  - Increase k-point density or adjust smearing width (e.g., Methfessel-Paxton or Gaussian).
  - Use advanced mixing techniques like Pulay or Anderson methods to accelerate convergence [15].

3.10.2.3 Memory or Runtime Limit Exceeded

- **Causes**: Insufficient computational resources for large systems.
- **Solutions**:
  - Reduce the system size or complexity (e.g., smaller supercell).
  - Optimize FFT grid and parallelization.
  - Request more memory or computational nodes in the HPC environment.

3.10.3 Post-processing

Post-processing involves extracting physical properties and interpreting the data generated by the DFT calculations.

3.10.3.1 Density of States (DOS)

- **Purpose**: Provides the distribution of electronic states at each energy level.
- **Key Insights**:

  - Identify the Fermi energy for metals.
  - Determine band gap size for semiconductors and insulators.

- **Tools**: Built-in DOS calculators in DFT codes.

3.10.3.2 Band Structure

- **Purpose**: Plots energy levels as a function of crystal momentum along high-symmetry k-points in the Brillouin zone.
- **Procedure**:

  - Perform a non-SCF calculation using a pre-converged density.

- **Insights**:

  - Identify band gaps, direct/indirect transitions, and band dispersions.
  - Note: DFT underestimates the band gaps. Thus, one may not get the correct band gap value.

3.10.3.3 Charge and Spin Density

- **Purpose**: Visualize bonding, electronic localization, and magnetism.
- **Tools**: Visualization software like VESTA or XcrysDen.

3.10.3.4 Forces and Stresses

Verify that forces and stresses are minimal after relaxation to ensure the stability of the system.

3.10.3.5 Derived Properties

Using the obtained eigenstates and eigenenergies, several physical properties of solid under study can be calculated. Various DFT packages are provided with the options to calculate these properties. Moreover, there are several other available codes, which are interfaced with DFT packages, that calculate different physical properties of solid under study. Some of the properties that can be calculated using the results of DFT calculations are-

- **Transport Properties**: Electrical conductivity (σ), electronic part of thermal conductivity (κ), Seebeck coefficient, electronic specific heat, electron and hole mobility in semiconductors, anomalous Hall Conductivity (driven by Berry curvature), and many more.
- **Optical Properties**: Frequency-dependent dielectric constant, refractive index, absorption coefficient, reflectivity spectra, response of materials to energy loss of electrons, optical excitation leading to current, etc.
- **Vibrational Properties**: Phonon modes and frequencies (with DFT + phonon calculations).
- **Magnetic Properties**: Spin polarization and magnetic moments.
- **Elastic Properties**: Bulk modulus, shear modulus, and Poisson´s ratio derived from stress-strain relationships.

3.10.3.6 Comparison with Experimental Data

Validate calculated properties such as lattice constants and band gaps against the available experimental results or theoretical works in literature.

3.10.4 Practical Tips for Post-processing

- Using Python scripts to process large datasets is highly recommended.
- For the proper analysis of the results, one must generate high-quality plots using Matplotlib, Gnuplot, Xmgrace, or OriginPro.
- Compare the obtained DOS and dispersion curves with the previously available data in literatures. Look for any incorrect band crossings or unexpected DOS contributions. For instance, correct band gap must be reflected from the DOS curve.
- Last but not the least, maintaining a well-organized records of input/output files and post-processing scripts is highly recommended for data reproducibility, whenever required.

By following above discussed suggestions, DFT calculations can produce useful results that may prove to be necessary stuffs for exploring the underlying physics of any given material.

References

1. E. Wimmer, et al., Full-potential self-consistent linearized-augmented-plane-wave method for calculating the electronic structure of molecules and surfaces: O_2 molecule. Phys. Rev. B **24**(2), 864 (1981)

2. F. Giustino, *Materials Modelling using Density Functional Theory*, 1st edn. (2014). ISBN 978–0–19–966243–2
3. N. Ashcroft, N. Mermin, *Solid State Physics (Cengage Learning)* (2011)
4. S. Cottenier, Density Functional Theory and the family of (L)APW-methods: a step-by-step introduction, 2002–2013, 2nd edn. ISBN 978-90-807215-1-7 (freely available at http://www.wien2k.at/reguser/textbooks)
5. O.K. Andersen, R.V. Kasowski, Electronic states as linear combinations of muffin-tin orbitals. Phys. Rev. B **4**(4), 1064 (1971)
6. O.K. Andersen, R.V. Kasowski, Electronic states as linear combinations of muffin-tin orbitals. Phys. Rev. B **4**(4), 1064 (1971)
7. R.M. Martin, *Electronic Structure Basic Theory and Practical Methods*, 2nd edn. (2020). ISBN 978-1-108-42990-0
8. M. Petersen, et al., Improving the efficiency of FP-LAPW calculations. Comput. Phys. Commun. *126*(3), 294–309 (2000)
9. G. Kresse, J. Furthmüller, Efficient iterative schemes for ab initio total-energy calculations using a plane-wave basis set. Phys. Rev. B **54**(16), 11169 (1996)
10. M. Novák, et al., Adaptive Anderson mixing for electronic structure calculations. Comput. Phys. Commun. 108865 (2023)
11. M. Winkelmann, et al., Kerker mixing scheme for self-consistent muffin-tin based all-electron electronic structure calculations. Phys. Rev. B **102**(19), 195138 (2020)
12. P. Blaha, et al., WIEN2k: an APW+ lo program for calculating the properties of solids. J. Chem. Phys. **152**(7) (2020)
13. T. Loucks, J.C. Slater, Augmented plane wave method: a guide to performing electronic structure calculations. Phys. Today **20**(11), 92–93 (1967)
14. J.P. Perdew, K. Burke, M. Ernzerhof, Phys. Rev. Lett. **77**, 3865 (1996); **78**, 1396(E) (1997)
15. R. Awada, C. Jérôme, C. Rosier, Anderson acceleration. Convergence analysis and applications to equilibrium chemistry. Appl. Numer. Math. (2024)
16. E. Wimmer, A.J. Freeman, Fundamentals of the electronic structure of surfaces. In: *Handbook of Surface Science*, vol. 2, pp. 1–91 (North-Holland, 2000)
17. http://exciting.wikidot.com/neon-understanding-species
18. https://docs.abinit.org/variables/basic/

Chapter 4
Software and Tools for Density Functional Theory

Pramesh Gunawardhana and Roshan Thotagamuge

Abstract Density functional theory represents a revolution in the description of electronic structure in materials, molecules, and complex systems. The aim of the present chapter is to introduce the reader to DFT software-indispensable tools to unlock the power of this theoretical framework. First, an overview will be given about some of the freely available packages—quantum ESPRESSO, OpenMX, Yambo, among others—and commercial ones like Gaussian, ADF, and VASP-together with their strengths, weaknesses, and applicability to material science and biological systems. Then, it will be shown in a tutorial how to perform a basic DFT calculation step by step: it will be shown how to define the system, how to take into account possible high-performance computing, how to choose functional and basis sets, and how to analyze the output. This tutorial will provide the reader with the background needed to use DFT codes for research purposes with the help of visualization tools. The chapter allows the reader to connect what has been shown so far to the actual practice and use of DFT software.

Keywords Density functional theory (DFT) · Computational chemistry · Gaussian

4.1 Introduction

DFT is a very powerful theoretical framework that has dramatically changed our understanding of electronic structure in materials, molecules, and complex systems. Using methods borrowed from quantum mechanics, DFT enables a wide range of physical and chemical phenomena to be modeled and predicted [1]. In practice,

P. Gunawardhana (✉)
Department of Nano Science Technology, Faculty of Technology, Wayamba University of Sri Lanka, Kuliyapitiya, Sri Lanka
e-mail: gunawardhanapramesh@gmail.com

R. Thotagamuge (✉)
Department of Optometry, Faculty of Allied Health Sciences, University of Sri Jayewardenepura, Nugegoda, Sri Lanka
e-mail: roshan@sjp.ac.lk

R. Thotagamuge (ed.), *Density Functional Theory*,
https://doi.org/10.1007/978-981-95-9745-1_4

the mathematical complication lying behind DFT often requires the use of specific software tools.

The chapter now deals with the software tools developed for DFT and is meant to unlock its power. We embark on a journey starting with an overview of popular DFT software. In the lectures of this section, some features of the freely available software packages—Quantum ESPRESSO, OpenMX, and Yambo—are discussed, together with those of the commercial software packages Gaussian, ADF, VASP, in order to familiarize you with a broad sample of existing software packages, for quite different needs and levels of expertise [2]. We will discuss strengths and functionality for some of the most widely used programs; then, we highlight their suitability for particular applications such as materials science applications and simulations of biological and chemical systems.

After this overview of the software, the basic use of such tools will be provided to you. In the next section, General Guide to Using One of These Tools a step-by-step approach will be provided to setup and execute a basic DFT calculation using one of the popular suites of programs. This tutorial will walk you through such a typical process: the definition of the system; then the choice of an appropriate High-Performance Computing tool in case large systems have to be investigated; the actual selection of a proper functional and basis set, eventually ending with the interpretation of the obtained results. This will be done supported by Visualization Tools for DFT Results [3].

To consolidate your learning of the practical issues in DFT calculations, the section will carry out an Example of a Simple DFT Calculation. This includes a realistic case study that will take you through the preparation and execution steps for the DFT calculation of a system. Dissecting the example step by step will give experience in the running of DFT software to solve practical problems.

At the end of the chapter, the reader will be ready to apply DFT software to help their own research. We will get the reader ready by taking them through the different options of the software, the choice of the tool for a given job, and how to carry on the DFT calculation, on an understandable road to discovery into the wonderful land of electronic structure.

This chapter connects the theoretical basis of DFT to its practical applications and gives an overview of the software tools, with enough practical information to provide you with the necessary skills to unlock the transformative potential of DFT in your scientific pursuits [4].

4.2 Overview of Popular DFT Software

This will be an introductory act to our journey: a bird's eye overview of the most popular DFT software packages, freeware, and commercially licensed. We will enter a fascinating world of free and commercially licensed packages. In the free-of-charge realm, we will introduce some powerhouses like Quantum ESPRESSO, OpenMX, and Yambo, each serving specific needs. As for the commercial solutions, we will take a closer look at the industry-leading offerings: Gaussian, ADF, and VASP. The next section will show, besides strengths and functionalities, the applicability of these programs to targeted applications. We are going to delve into details on how some programs are absolutely stellar for material science simulations, while others are fantastically good at handling biological and chemical systems. At the end of this section, you will be well-versed in negotiating the software landscape and in selecting the appropriate tool for your DFT undertaking.

4.2.1 Freely Available DFT Software Packages: Open-Source Powerhouses for Diverse Applications

Our journey begins with a spotlight on the exciting world of freely available DFT software packages. These open-source options provide a powerful and accessible entry point for researchers and students alike. Here, we'll encounter renowned names like Quantum ESPRESSO, known for its efficiency and scalability. We'll also explore OpenMX, a versatile tool particularly well-suited for simulations involving electronic transport phenomena. Additionally, Yambo will be introduced, a software package specifically designed for calculations involving electronic excitations in materials. This section will delve into the unique strengths and functionalities of these freely available programs, empowering you to make informed decisions when selecting the most suitable tool for your specific DFT calculations.

Quantum Espresso

Quantum ESPRESSO is a giant among freely available DFT packages. Renowned for efficiency and scalability, and by the breadth of its functionalities, it steadily became a cornerstone in every good researcher's and student's training curriculum in a variety of scientific fields. This section shall focus on key strengths that render Quantum ESPRESSO an attractive choice for electronic structure simulations across a broad range [5].

Released and developed under an Open-Source license, Quantum ESPRESSO is an open, transparent, and collaborative environment. It allows continuous improvement through active participation from its developer community and ensures that the software is always on the edge of DFT advancement [6]. Other than being open-source, one of the strong points is impressive efficiency in handling complex systems. Using plane-wave basis sets and pseudopotentials, it can perform the most demanding calculations with speed and accuracy.

But that is not all: Quantum ESPRESSO is also fantastically scalable. It leverages parallelization techniques to scale calculations easily across many processors, making it the ideal choice for any HPC environment. The scalability further enables researchers to attack tricky simulations of big molecules and materials, therefore enabling groundbreaking science [7].

Besides the above-mentioned key strengths, Quantum ESPRESSO provides a very rich set of functionalities for the many and diverse DFT applications: it allows a wide variety of functionals to be used, providing this way the possibility to tailor the calculation for a particular problem; it includes tools for structural optimization, molecular dynamics simulation, and response properties calculations. Those who will go into the detail of material properties will be able to use Quantum ESPRESSO to calculate band structures, Fermi surfaces, or electron-phonon interactions. Another key feature is that it interacts well with various post-processing packages in order for the researcher to visualize, and further analyze in much greater detail, their DFT result [6].

We notice, incidentally, that Quantum ESPRESSO is more oriented toward the class of advanced users who are comfortable with text input files. While GUIs are existing, in fact, most functionalities in Quantum ESPRESSO require user-supplied input files [8]. The steep learning curve is well compensated, however, by the extensive documentation and the active user community.

In the next sections, we will introduce the user experience of Quantum ESPRESSO. We go into more detail concerning its functionalities and show its applicability to a broad range of uses under the framework of DFT calculations.

Quantum ESPRESSO is a powerhouse among open-source DFT codes. Because of its efficiency, scalability, and the extent of its features, it has widespread use among both researchers and students across various scientific disciplines. The user experience is somewhat a two-edged sword: while having pros, it also carries some overhead in learning [9].

Probably the most striking fact about Quantum ESPRESSO is that it is free software. That means that its source code is freely available for anybody to view, modify, and contribute to. This offers a collaborative environment whereby an exceptionally enthusiastic developer community keeps developing this software for fresh DFT advances. Another implication of free software is that there are no licensing costs [10].

Quantum ESPRESSO really works wonders when it comes to complex calculations. Using plane-wave basis sets and pseudopotentials, it does the most resource-intensive simulations faster and more accurately than anything else out there. Just envision that kind of world where "an eternity of intricate calculation" doesn't actually take an eternity [11].

But the magic does not stop here: Quantum ESPRESSO is also strikingly scalable. By applying the principles of parallelization, it distributes calculations with ease over many processors, becoming the perfect candidate to run on HPCs. By using scalability, this enables researchers to perform intricate simulations of large molecules and materials and thus further enhance scientific knowledge.

Quantum ESPRESSO goes much beyond basic calculations and provides a rich palette of functionalities for very different applications of DFT. Software packages that form such a broad-based toolbox might be expected to support a large variety of different functionals the different tools in the box with which one can tailor the calculations to particular problems and provide functionalities for the optimization of molecular structures, molecular dynamics simulations, and the calculation of response properties of materials to external stimuli such as electric fields or light [12].

Capabilities of Quantum ESPRESSO range from computing properties that include band structure-the energy of the electrons in a solid-and Fermi surfaces, which give a view of the distribution of electrons in momentum space, to electron-phonon interaction, describing how electrons interact with atomic vibrations [13]. Besides that, it interfaces with a whole suite of post-processing packages that make detailed visualization and analysis of DFT results possible.

While Quantum ESPRESSO is indeed a powerful platform, there are some considerations in the user experience. Unlike some software with very user-friendly graphical interfaces, Quantum ESPRESSO mainly operates through text input files. That means users will have to feel comfortable scripting and editing code in order to define the system, choose calculation parameters, and interpret the results [14].

Quantum ESPRESSO is like a very powerful sports car; amazingly powerful and flexible, but at the same time, it needs some technical expertise to fully master. The learning curve might appear steep at first regarding the input file format and syntax, but huge rewards are gained in return: amazingly accurate computational results and the capability to deal with incredibly complex problems. Most of the challenges that come with the use of Quantum ESPRESSO derive from its great flexibility and depth of capabilities. Mastering Quantum ESPRESSO therefore requires that users take the time to learn about the format and syntax of the input file, even though there are GUIs available, and most core operations are still by and large controlled by user-generated input files. This, in turn, means a moderate knowledge of command-line operations will often be required. Sometimes there is even an overabundance of available options and parameters, possibly leading to confusion for the novice user in trying to choose the most appropriate settings for a particular calculation. These challenges can be surmounted through further dedicated learning and consideration of the available resources and tutorials.

The good news is that there are resources available to mitigate these challenges and ease the learning curve. Extensive documentation and tutorials accompany Quantum ESPRESSO, providing valuable guidance for users. Furthermore, an active online community of users fosters a supportive environment where users can seek help, share knowledge, and troubleshoot issues together. To further simplify the process, graphical user interfaces (GUIs) like PWgui can assist in generating input files, particularly beneficial for beginners. These resources significantly enhance the user experience and make the learning process more accessible [15].

In all, Quantum ESPRESSO is a powerful tool for advanced users who do not have any problems with text input. Due to its open-source nature, efficiency, and functionalities packed within, it becomes equally useful for researchers. However, this involves a steep learning curve in writing the input files yourself, so one has to be in a position where they can invest time in learning-with the help of documentation and support from the community. Once you put that little effort in, the immense power of Quantum ESPRESSO will be at your fingertips and will open an entirely new portal into this fascinating world of DFT calculations.

OpenMX

OpenMX is one of the most flexible free packages available within the DFT realm. It has been recognized to be user-friendly and to have high efficiency, mainly for calculations related to electronic structure and transport phenomena. This section elaborates on the core strengths that OpenMX embodies and is thus a great option for anyone be it a researcher or a student who has interests in this particular field [16].

OpenMX places a high level of importance on user experience by providing various ways to input parameters. Whereas several software packages depend on an intricate, single text-based input file, OpenMX provides alternatives, such as script-based input for advanced users who can leverage scripting languages like Python for streamlined workflows. In addition, the new matrix-based input provides the possibility to define calculations in terms of matrices, as many researchers naturally think about and describe their systems. Thirdly, although limited compared to the core capabilities, the available GUIs ease the creation of input files for beginners and thus lower the learning curve for using the software. The flexibility in input methods makes OpenMX more user-friendly and considerably broadens its field of applications to a wide class of users [17].

OpenMX is unique in its capability to simulate electronic transport phenomena, such as very important processes like electrical conductivity and thermal transport in materials, using powerful computational techniques such as the Non-Equilibrium Green's Function method. This allows for the precise simulation of current flow through nanostructures and molecular devices. Besides, OpenMX uses the Landauer formalism, a well-established theoretical framework for calculating electrical conductance in a wide range of systems. These advanced functionalities make OpenMX an indispensable tool for researchers working in the fields of molecular electronics, spintronics, and optoelectronic devices, where understanding and controlling electron transport is of paramount importance [16, 17].

While transport phenomena are a particular strength of OpenMX, it offers a broader range of functionalities for general DFT calculations. Like Quantum ESPRESSO, it has quite a few different functionals and tools for tasks such as; Structural optimization, Molecular dynamics simulations and Calculations of response properties.

Such versatility permits OpenMX to treat an even broader class of problems than just transport phenomena [18]. In addition to basic strengths discussed above, the following are the features of OpenMX, providing extra power. OpenMX uses localized atomic orbitals as basis sets, which may be handy in calculations for large molecules or periodic systems. The calculation efficiency translates to a much faster computation time compared to other DFT software using plane-wave basis sets. OpenMX is capable of mixing different types of basis sets in a single calculation. Such flexibility would be useful in treating a whole system where certain regions need detailed treatment. OpenMX is integrated with other post-processing tools, which provides researchers with the capability to visualize and analyze in detail the results obtained from DFT calculations. Such tools may also be useful in extracting vital information from calculations, like electronic band structures, densities of states, and charge distributions [19].

It should be noted that, even though OpenMX has a GUI for script-based and matrix-based input, a certain level of familiarity by the user is needed regarding the concepts in DFT and scripting languages in order to delve into more complex functionalities. However, it is relatively easy to obtain support from the active user community, and detailed documentation is provided for beginners in the software. In general, OpenMX provides a user-friendly and versatile open-source platform for DFT calculations. It does very well in the simulation of electronic structure and transport phenomena, so it's a very important tool for researchers in those fields [18].

Yambo

While Quantum ESPRESSO and OpenMX are the best in a few features of DFT calculations, Yambo takes an alternative route. This free and open-source software package concentrates on a very critical aspect mostly overlooked by conventional DFT calculations: electronic excitations.

Standard DFT calculations are mostly carried out for the ground state, which is the most stable configuration of any system. More interestingly, though, most materials and molecules have properties that arise from their excited states. Excited states play a vital role in phenomena such as light absorption, luminescence, and electrical conductivity [20].

Yambo will fill the gap by providing a robust toolkit to compute the excited-state properties of solids and molecules using the Many-Body Perturbation Theory (MBPT) and the time-dependent density functional theory (TDDFT). The advanced techniques give an opportunity to researchers to study the electronic behavior of materials when excited by light absorption or Evison injection.

This is how Yambo differs from a standard DFT code. Yambo goes further and computes the properties of excited electronic states, not only the ground state. It uses state-of-the-art techniques of MBPT and TDDFT to investigate the electronic response of the material to some external stimulus. Simulation of light absorption and emission makes Yambo an extremely useful tool for researchers in the field of photovoltaics, light-emitting diodes (LEDs), and solar energy conversion [21].

It also has an open-source philosophy like Quantum ESPRESSO and OpenMX, which ensures the transparency of the software and collaboration among the scientific community; this makes this powerful tool more accessible to many researchers and institutions.

Apart from the basic excited state calculations, Yambo has some features which further enhance its capabilities. Yambo has also been interfaced with well-established ground-state DFT codes, like Quantum ESPRESSO (QE), to take advantage of existing ground-state calculations from QE as input to excited-state calculations in Yambo to provide a smooth workflow. Yambo goes beyond simulating light absorption; actually, it deals with the whole range of excitations, including electron injection, which is crucial to understand phenomena such as photocatalysis and solar cell operation. Starting from excited state information, Yambo can calculate various optical properties of the materials, including absorption spectra and refraction coefficients, which involve photoluminescence spectra, critical for designing materials with tailored optoelectronic applications in mind [21, 22].

One should keep in mind that Yambo is geared more toward the experienced user with a good background in concepts of DFT and many-body theory. Indeed, calculations may involve fairly complex input files, and one might need advanced knowledge to interpret results from the calculations. Still, the Yambo community and available tutorials do help bring guidance to the researcher venturing into this exciting world of excited-state calculations. In summary, Yambo is a unique and powerful tool that has been provided to the user for exploring the exciting world of electronic excitations in solids and molecules. It offers researchers the possibility of going beyond standard DFT calculations with advanced techniques and studying complex properties of materials in a wide energy range of excitations [23].

4.2.2 Commercial DFT Software Packages: A Closer Look at Industry-Leading Solutions

While open-source software packages such as Quantum ESPRESSO, OpenMX, and Yambo bring powerful capabilities, the commercial DFT software packages generally offer a more complete and user-friendly experience. The industry-leading solutions, now very well developed and optimized and available for an extremely broad range of computational chemistry and materials science applications, include Gaussian, VASP, and ADF.

While offering advanced functionality, intuitive interfaces, and strong technical support, commercial packages are targeted toward researchers who need streamlined workflows and access to the most advanced methods available. We review some of these main features and advantages of prominent commercial DFT software packages in the following sections [24].

Gaussian

Gaussian has been a genuinely pioneering software package in the area of computational chemistry; it has played a quite substantial role in pushing forward our knowledge of molecular systems. Since its inception, Gaussian has grown into a strong and quite versatile platform that now hosts an overwhelming number of computational methods for investigating the properties and behaviors of molecules. The key features, applications, and impact of Gaussian on chemical research have been discussed in the following section [25].

Such strength lies with Gaussian in the ability to provide good predictions with reliability for a wide chemical property range. Gaussian, from its geometries and energies to spectroscopic data and mechanisms of reactions, offers a toolkit that can investigate the complex chemical phenomena. Due to its user-friendly interface and large documentation part, it has been widely adopted in different scientific disciplines and becomes indispensable in academia and industry.

Gaussian has played a central role in the development and application of density functional theory (DFT) in computational chemistry. It contains a large variety of DFT functionals, enabling the accurate determination of results for electronic structures, energies, and properties of molecules and materials. The implementation of DFT provided in Gaussian has been of utmost importance in leading to improved understanding in chemical bonding, reactivity, and spectroscopy. Additionally, Gaussian also offers hybrid methods, combining DFT with wavefunction-based approaches in a powerful framework for the study of complex chemical systems [25, 26].

In the following sections, we will discuss more detailed applications of Gaussian in structure optimization, electronic structure calculations, reaction kinetics, and spectroscopy. Together with an explanation of the core functions of Gaussian and its contributions to DFT, a reader should be able to get an idea of the capabilities of this software and its importance in modern chemical research [27].

Gaussian excels in the calculation of molecular geometries. Structure optimization is the process of searching for an arrangement of atoms that corresponds to the lowest energy state. Gaussian can predict very good values for molecular conformation, bond length, and bond angles by minimizing the system's energy with respect to atomic coordinates. Such information is very important for the estimation of stability, reactivity, and spectroscopic properties of molecules.

The ability to calculate the electronic structure forms the basis of computational chemistry, since it provides information about properties and reactivity. Gaussian provides a very extensive set of methods for electronic structure calculations: Hartree–Fock, density functional theory, and post-Hartree–Fock methods. These allow for descriptions of molecular orbitals, energies, dipole moments, and other electronic properties [28]. The insights obtained from electronic structures contribute to explaining chemical bonding, reactivity, and spectroscopic behavior.

Gaussian allows the investigation of chemical reactions by computing potential energy surfaces (PESs). These surfaces present the energy of a system as a function of atomic coordinates, and they can be used to learn about reaction pathways, transition states, and activation energies. Knowledge of PESs enables the investigation of reaction mechanisms, the prediction of reaction rates, and the determination of kinetic parameters. These are things that Gaussian can do in order to understand chemical processes and design efficient catalysts [29].

Spectroscopy is a very useful experimental method for the investigation of molecular structures and properties. Gaussian provides a complete set of tools for simulating different spectroscopic experiments, including infrared (IR), Raman, nuclear magnetic resonance (NMR), and ultraviolet-visible (UV-Vis) spectroscopy. Comparing the calculated and experimental spectra allows validation of the theoretical models, assignments of the spectral features, and derivation of structural information. The spectroscopic capabilities in Gaussian really come in handy when characterizing molecules and working out reaction mechanisms [30].

These core functionalities combined empower Gaussian to address an enormous variety of chemical problems, from the design of new drugs to the development of advanced materials.

VASP (Vienna Ab-initio Simulation Package)

VASP stands for the Vienna Ab-initio Simulation Package. It is among the most exclusive commercial software packages available in atomic-scale simulation and calculation. Given its extraordinary precision, effectiveness, and flexibility, this would have to be rated as an extremely good tool in the essential field of material sciences and condensed matter physics.

At its very core, VASP implements DFT as the fundamental platform. It is a more computationally palatable means of approximating the solution to the quantum many-body problem, whereby the researchers can study the electronic structure of materials with great precision. Implementation of DFT in VASP evolves much further than the mere inclusion of state-of-the-art methodologies, such as hybrid functionals and perturbation theory, which would be more apt to describe subtle electronic correlations and exchange interactions [31].

Apart from ground-state DFT, the range of computational techniques available in VASP is enormous. It grants the possibility to make use of an entire range of exchange-correlation functionals: LDA, GGA, meta-GGA, and hybrid functionals. Such flexibility in the use of exchange-correlation functions offers the user free choices in most realistic calculations, fitted to the demands of each system [32]. VASP supports various basis sets, including plane-wave and projector-augmented wave methods, which offer flexibility toward different computational needs.

Another powerful ingredient for the high acceptance of VASP is its computational efficiency. The code effectively exploits parallel computing architectures, which enables the DFT simulation of challenging problems on modern HPC systems. This becomes an important ingredient when one wants to study large-scale material systems or explore extensive parameter spaces. Some specific capabilities regarding certain aspects of the Materials Sciences and Condensed Matter Physics that can be done by using VASP will be underlined below, namely, in respect to the following: structural optimization, electronic structure calculation, and molecular dynamics.

VASP excels in determining the equilibrium geometry of materials. Through a process known as structural relaxation, VASP iteratively adjusts atomic positions and cell parameters to minimize the system's total energy. All this turns out to be important in predicting the structures of crystals, lattice parameters, and atomic positions, forming the very necessary basis to understand material properties. The efficiency of VASP for big systems, with additional functionality for imposing diverse constraints, makes this tool absolutely irreplaceable in investigating complex materials and their structural transformations [32, 33].

VASP thus gives a firm basis for the calculation of electronic properties like electronic band structures, density of states, and charge densities from a Kohn-Sham solution within the DFT framework. The latter will provide insight into the nature of conductivity, magnetism, or optical behavior. It also gives effective masses, Fermi surfaces, and dielectric functions pertinent to charge carrier transport and optical response. Beyond these static properties, VASP is also capable of modeling material dynamics via MD simulations. Solution of Newton's equations of motion for the atoms, with the interatomic forces obtained from DFT, is used by VASP in order to determine the atomic trajectories the system follows with time. In that respect, thermal properties, phase transitions, diffusion, and other dynamic phenomena can be studied with this approach.

The ability to account for sophisticated interactions and large-scale simulations finally turn VASP into an ultimate tool in investigations on the behavior of materials in different conditions [34]. These, together with the effective and precise treatment of VASP, provide the core functionalities responsible for setting it as a standard in materials science and condensed matter physics.

Amsterdam Density Functional (ADF)

ADF is a very mature commercial package, fully recognized by the communities of computational chemistry and materials science. Developed by SCM, Scientific Computing and Modeling, ADF allows a whole range of activities related to the study of electronic structure and properties of molecules, solids, and surfaces. The base ideas of ADF grow from the variation of density functional theory as the heart of its computational tool, able to provide very accurate and reliable predictions across a wide range of chemical and physical phenomena [35].

Beyond DFT, ADF embeds an extremely large number of state-of-the-art computational methodologies, like wavefunction-based approaches, time-dependent DFT, and the embedded cluster method. Immediately, this allows the modeling of very complex systems, investigating various properties with high accuracy. ADF is very friendly to the user, thanks to its large library of functionalities, of great value for academic and industrial research. That is a great strength of ADF: handling different chemical systems, ranging from small molecules to extended solids and interfaces, describing with great ease the basically covalent, ionic, and metallic bonding situation. In ADF, a number of choices concerning basis set flexibility, exchange-correlation functionals, and solvation models are allowed.

A selection of the current capabilities of ADF, in the area of computational chemistry and materials science is given in the following sections: Molecular and Electronic Structure Calculations; Reactivity Studies; Spectroscopy [36, 37]. ADF is quite strong concerning molecular and electronic structure description. Due to the enormous library concerning DFT functionals and basis sets included, the calculations can be made very system-specific. Important features of the ADF are; determining the most stable molecular conformation through energy minimization, Infrared and Raman vibrational spectra predictions with a view to identifying and characterizing molecules, molecular orbital and electron density calculations including other electronic properties aimed at understanding bonding and reactivity and Investigating electronic excitations through TDDFT to study spectroscopy and photochemistry [38].

ADF provides a complete set of tools for the study of chemical reactivity, from the study of reaction mechanisms by following reaction paths, locating transition states, calculating activation energies, and giving kinetic interpretations to building potential energy surfaces in order to identify stable intermediates and understand reaction profiles. Besides, ADF allows calculating reactivity indices such as Fukui functions, electrophilicity indices providing useful insight to the characterization of reactive sites and problems of selectivity. This solver includes in an essential manner solvent effects, so that solution-phase reactions may be modelled and its effect on the reactivity may be studied.

On top of investigations of reactivity, ADF is also a real tool-box for theoretical spectroscopy. That includes the simulation of electronic transitions and absorption spectra so important in UV-Vis spectroscopy, greatly contributing to the characterization and identification of molecules. Besides, ADF allows the calculation of vibrational spectra, including IR and Raman, which makes the interpretation of experimental data and structural elucidation easier. Furthermore, it can predict NMR chemical shifts and coupling constants, aiding in structural determination and analyses. Further, ADF contains tools for modeling EPR spectra to study the properties of radical species.

In this, the core functionalities of ADF, its application to DFT-based simulations, will be discussed, helping readers find out its potential in view of advances in chemical and materials research [39]. The previous sections identified some of the commercial and free DFT packages, and gave an idea about their relative strengths and application areas in computational chemistry and materials science. One can comfortably solve a wide range of scientific challenges with much ease by blending commercial and free DFT software together. It is this synergy—the mutual intertwining of theory and computation—which DFT is an example of that has opened up whole new avenues to chemistry, physics, and materials science.

As computational capability and algorithms continue to develop, so does DFT simulation art in sophistication and accuracy that can be foreseen in the future. Integration with machine learning, artificial intelligence, and high-performance computing will further enhance the capability of such software packages in enabling researchers to handle increasing complexity in challenges and frontiers of knowledge.

4.3 General Guide to Using Gaussian

Gaussian is one of the most important computational chemistry software; it allows a really powerful tool to model and simulate almost all kinds of chemical systems. This section will present a step-by-step description about using the Gaussian software: building an input file, running the calculation, and then analysis of the output. Understanding how the basics and flow work enables the user to take the software to solve many key chemical problems.

Whereas Gaussian offers a very rich set of computational methods and capabilities, this tutorial will illustrate only how to perform the basic manipulations necessary to execute a typical DFT calculation using Gaussian; these will suffice to provide the reader with a starting point from which to move on to more specialized uses [40].

4.3.1 The Job Processing Window

The Job Processing window is the home from which Gaussian jobs are managed and run, and also from which their output is reviewed. The principal elements within this are summarized in Fig. 1.

The rest of this chapter describes the menus and buttons in this window.

The File Menu

The File menu provides means to create and retrieve Gaussian input files, and to specify program preferences.

1. **New**: Create new Gaussian input (residing only in memory until it is explicitly saved to disk).
2. **Open**: Open an existing Gaussian input file. The extension of a Gaussian input file is GJF. "The Open…" menu item may also be used to load an existing batch control file. The batch facility is described later in this section. Finally, it may be used to open a PDB file for conversion (this process is discussed later).
3. **Modify**: Edit existing input, using the Existing File Job Edit window.
4. Preferences: Specify Gaussian preferences. Preferences are detailed in a separate section later in this document.
5. **Exit**: Quit Gaussian. You will be prompted whether to save any unsaved new or modified input files as well as any unsaved changes to the preferences.

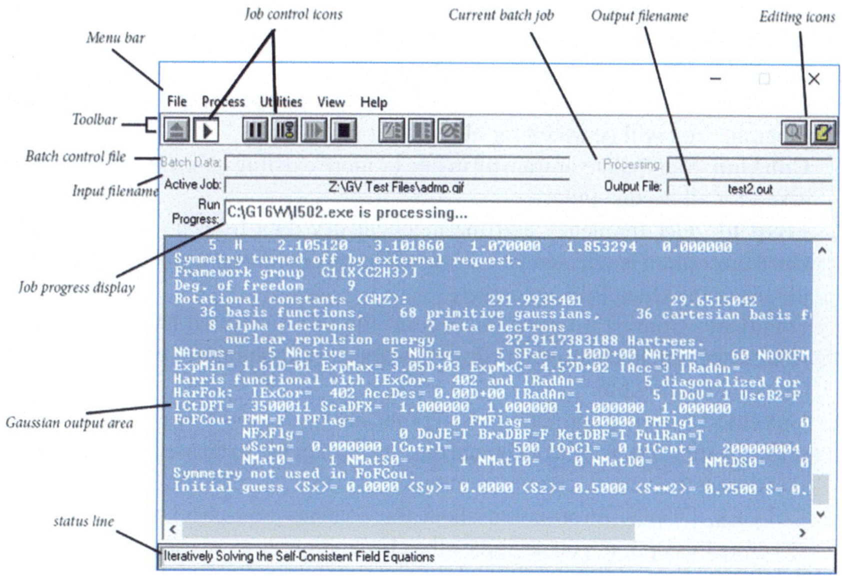

Fig. 1 Job processing window

The Process Menu

The Process menu lets you control running jobs. All of its entries have corresponding icons in the Job Processing window (see below).

1. **Start Processing**: Start processing the currently loaded input.
2. **Pause**: Immediately suspend the currently running job.
3. **Pause → Next Link**: Suspend execution of the currently executing job after it completes the current link.
4. **Resume**: Restart execution of a paused job.
5. **Kill Job**: Immediately terminate the currently executing job. If a batch is running, the next job in the batch (batches are formally defined later in this section) will start executing (unless the End Batch Run on Error preference is set).
6. **End Batch**: Terminate the execution of the current batch when the current job completes.
7. **Kill Batch**: Immediately terminate the currently executing job and end batch processing without executing any further jobs.

The Utilities Menu

The Utilities menu provides access to the batch and file conversion facilities, and to other utilities supplied with Gaussian.

1. **Edit Batch List**: Edit the currently loaded batch control file, via the Edit Batch List window. If no batch control file is loaded, then a new batch list is created and any currently loaded input is erased from memory.
2. **NewZMat**: Launch the NewZMat utility to convert files. Once this option is selected the user is asked to select a file to be converted via the Open File dialog box.
3. **CubeGen**: You can generate a cube file for viewing with a visualization program. You will be asked for all relevant information.
4. **CubMan**: Manipulate or transform one or more existing cube files. You will be asked for all needed input.
5. **FreqChk**: Get frequency and thermochemistry data from a checkpoint file. Once this option is selected you are asked to specify the name of the checkpoint file using the Open File dialog box.
6. **FormChk**: Convert a binary checkpoint file to a formatted (ASCII) version. After selecting this option, you designate the checkpoint file to be used with the Open File dialog box.
7. **UnFchk**: Convert a formatted checkpoint file back to its binary format. After selecting this option, you designate the checkpoint file to be used with the Open File dialog box.
8. **ChkChk**: Display information about the contents of a checkpoint file. After selecting this option, you designate the checkpoint file to be used with the Open File dialog box.

9. **C8609**: Convert checkpoint files between the current version and those for previous versions of Gaussian for Windows. After selecting this option, you designate the checkpoint file to be used with the Open File dialog box.
10. **External PDB Viewer**: View the current molecular structure with an external PDB viewing program. The program to use is specified in the preferences.

The View Menu

The View menu controls the appearance of the window, and allows you to invoke an external text editor. The default settings of the various display options may also be controlled via preferences. The editing options also have icon equivalents:

1. **Toolbar**: Toggles the display of the toolbar portion of the window. If the toolbar is visible, then this item is checked.
2. **Process Output**: Toggles on/off the displaying of the Output Display part of the window. If the Output Display part is displayed then this item is checked.
3. **Status Bar**: Toggles on/off the display of the status bar portion of the window which provides a brief description of the active pull-down menu item. If a status bar is present this item is checked.
4. **Editor**: open the external editor (which one opens depends on the setting).
5. **Editor → Output File**: Run the external editor on the current output file. Note that a job that is running needs to have been stopped before the editor is invoked on its output file.

Icons

The following icons are available in the Job Processing window—the similar menu item is listed below in parentheses:

 Start the execution of the current input or batch file. (Process: Run).

 Abort job immediately. (Process: Pause).

 Stop After Link Stop after the link in question. (Process: Pause → Next Link).

 Restart a suspended job (Process: Resume).

 Kill currently running job (Process: Kill Job).

 Edit the currently loaded batch control file, or create a new one if none is currently loaded (Utilities: Edit Batch List).

 Kill the running batch after the currently running job finishes execution (Process: End Batch).

 Immediately Abort the current batch, killing the current job (Process: Kill Batch).

 Open external editor open (View: Editor).

 Edit current output file. Remember to stop a running job first! (View: Editor → Output File).

Drag-and-Drop File Execution

Gaussian jobs can also be submitted by a convenient drag-and-drop method of execution. Move to the directory containing the desired input files, select the input file(s), drag onto the icon for Gaussian (if the program is minimized) or directly into the Job Processing window if it's open. If more than one file has been selected, a batch file named "Drop List. BCF" will be created containing the list of input files in the current directory. To start processing, "Begin Processing" can be selected from the "Process" menu (or via the icon). To simplify the user's workflow, Gaussian can be configured to automatically initiate the processing of the files dropped into the application window, easing submission for the user who submits calculations frequently [41–43].

We will go in-depth into how to create input files, submit jobs, and interpret the outputs with examples and explanations where necessary. This would also help both the beginning and advanced user be confident in navigating within the Gaussian environment and extracting meaningful insight from the calculations. Let's begin by exploring the structure of a Gaussian input file.

4.3.2 Understanding the Gaussian Input File for DFT Calculations

These parameters are very important for performing accurate and efficient DFT simulations using a Gaussian input file. Here, the major components are described in detail.

Route Section

Functionals

Whenever results of good accuracy are to be obtained, an important issue is the choice of the DFT functional. Among the commonly used functionals, one distinguishes the so-called Local Density Approximation, simple but usually less accurate in molecular applications, and Generalized Gradient Approximation, where the information on the gradient of the electron density is included, yielding more accurate results than the LDA functionals. Meta-GGAs improve the accuracy due to the kinetic energy density. Hybrid functionals represent a balance between accuracy and computational cost by mixing DFT with Hartree–Fock exchange.

Basis Sets

Select appropriate basis sets to describe the atomic orbitals. Common choices are Pople-style basis sets (6-31G(d), 6-311G(d,p), etc.) and Correlation-consistent basis sets (cc-pVDZ, cc-pVTZ, etc.), known for their systematic convergence properties.

Grid

Choose numerical integration grid for DFT calculations. Increasing grid density generally leads to increased accuracy but raises the computational cost. Standard choices of grids are Int = FineGrid (A standard grid for most calculations) and Int = UltraFineGrid (A denser grid for higher accuracy but increased computational cost).

SCF

Adjust SCF convergence criteria using keywords like scf = tight or scf = verytight for stricter convergence.

CheckPoint

Specify a checkpoint file to save intermediate results for possible future use.

POP

Request population analysis methods like Mulliken or Natural Population Analysis (NPA) to obtain atomic charges.

TDDFT

Enable Time-Dependent DFT calculations for studying excited states.

```
#P B3LYP/6-311G(d,p) opt freq pop=NPA scf=tight

Water Molecule

0 1
O 0.0 0.0 0.0
H 0.0 0.0 1.0
H 0.0 1.0 0.0
```

Fig. 2 Example input file

Constraints

Use constraints to fix particular atomic coordinates or bond lengths.

Example Input File for a DFT Calculation

The following input file describes a geometry optimization with a frequency calculation for a water molecule using the B3LYP functional with the 6-311G (d,p) basis set. It also asks for a population analysis and has a tight SCF convergence criterion (Fig. 2).

Careful choice of these parameters can make the Gaussian calculations tailor-made for the user's research subject and produce accurate and reliable results [44–46].

4.3.3 *Job Submission for DFT Calculations in Gaussian*

Once the input file has been generated, the next step is to submit the job to Gaussian software for execution. The submission process may vary in minor detail depending upon the computational environment; however, the general steps are as follows:

1. Navigate to the Gaussian executable: Locate the Gaussian executable file on your system. It is typically named g09 or g16 for versions 9 and 16, respectively.
2. Open a terminal or command prompt: Launch a terminal or command prompt window.
3. Change directory: Use the cd command to navigate to the directory where your input file is located.
4. Run Gaussian: Execute the Gaussian executable and specify the input file as an argument.
5. Monitor the calculation: The Gaussian software will process the input file and begin the calculation. You can monitor the progress in the terminal window.
6. Output file generation: Upon completion, Gaussian will generate an output file with the same name as the input file but with a ".log" extension.

Job Submission Options

Gaussian provides several ways to submit a job depending on the computational environment and/or desired features: Run Gaussian directly from the terminal or command prompt (Interactive mode), Submit the job to a batch queueing system for execution on a cluster or supercomputer (Batch system) and Run Gaussian on a remote server using tools like SSH or a job submission system (Remote execution) [46].

4.3.4 *Output Interpretation for DFT Calculations in Gaussian*

The output file from Gaussian for DFT calculations includes a great deal of information pertaining to the electronic structure, properties, and reactivity of the system under study. Careful analysis of this output is needed in order to extract meaningful insights and validate the results.

Key Components of the Output File

A standard Gaussian output of DFT calculation is very informative. It begins with a header section with general information such as the input file name, date, and calculation time. The route section gives the input-calculation parameters defined: the functional, basis set, and type of calculation requested, such as geometry optimization, frequency calculation. This will include the optimized atomic coordinates and the final energy if geometry optimization has been performed. The SCF convergence section of the output also indicates whether the self-consistent field procedure has converged successfully.

Among the more significant data, one can obtain the total electronic and nuclear energy of the system. The output file will also contain useful information such as Mulliken charges that give information about the electronic distribution inside the molecule. If asked for this output may also contain the DOS plot—a graph showing the electronic energy level distribution. For periodic systems, it is possible to compute and plot the band structure.

Also, if calculations of vibrational frequencies have been performed, the output file will contain vibrational frequencies and the corresponding normal modes. Finally, if TDDFT calculations have been conducted, the output contains transition energies and oscillator strengths of excited states that will be useful for the description of spectroscopic properties [46–48].

Analyzing the Output

A standard Gaussian output file has much to provide when it comes to crucial information required in the analysis of DFT calculation results. To start with, it verifies that the SCF procedure has effectively converged—a highly significant step since it then ensures that the electronic structure has reached a stable state. Second, it provides the total energy of the system, which is applied in analyses concerning the stability of systems as well as in comparisons among conformations or calculations. This also provides, in an output file, more useful information concerning the electronic structure, such as the Mulliken charges giving information on the distribution of electron density within the molecule. The DOS plot, if calculated, gives a representation of the distribution of electronic energy levels and thereby helps in identifying the occupied and unoccupied states. The band structure of the periodic systems is calculated and plotted, providing crucial insight related to the electronic conductivity of the material, along with other key features such as band gaps or metallicity.

The output file would print out the vibrational frequencies and corresponding normal modes, which can be directly compared with experiments if vibrational frequency calculations have been carried out. If TDDFT calculations have been performed, the output gives some interesting information about excited states, like transition energies and oscillator strengths, playing an important role for determining the system response to light or other external perturbations [49].

Visualization Tools

DFT calculations yield a multitude of results that are, by no means, self-explanatory. Part of the analysis process necessarily involves visualization of the results. Special capabilities for visualization of molecular structures, electron density, and molecular orbitals of output from Gaussian are incorporated into the GaussView, Avogadro, and VMD packages.

GaussView, designed specifically for Gaussian, provides an integrated combination and a complete set of visualization features. Avogadro, an open-source tool, is versatile and extendable with plugins. VMD was originally developed for molecular dynamics but is also competent for the effective visualization of DFT calculations, especially for large systems [50].

Key visualization techniques include molecular structures, electron density maps, molecular orbitals, electrostatic potential surfaces, and vibrational modes. It enables the researcher to further understand, from these visualization tools, results obtained from Gaussian calculations and make informed interpretations of such findings [51].

4.4 Example of A Simple DFT Calculation

Herein, the reader will be taken through a practical application of performing a DFT calculation for Gaussian software. The process entails setting up an input file of the calculation, running it, and analyzing its output. This example will provide you with first-hand experience in the application of DFT calculations and effectively build a strong foundation for handling more complex systems.

The example to be worked will include a simple molecule, and a simple DFT functional/basis set. We will be targeting a geometry optimization/frequency calculation for the main steps of a DFT calculation [52].

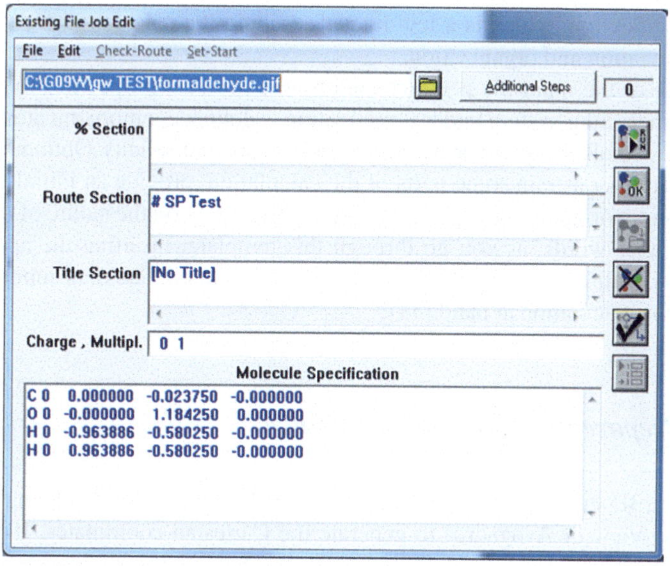

Fig. 3 Input file

This example will highlight the main features that are involved in carrying out a DFT calculation using Gaussian software; it starts with the most crucial preparation of the input file. The input file has to be prepared with care to accurately and succinctly describe the system being studied. At the end, the prepared job will be submitted for execution to Gaussian. Finally, we will explain in detail the very important step of output analysis, i.e., the interpretation of the main results obtained from the calculation: optimized geometry of the molecule, its total energy, and vibrational frequencies. This will be achieved step by step to provide a practical understanding of the entire DFT calculation process.

From this, you will learn how to apply concepts from earlier sections to one real example and get practical experience on the use of Gaussian with DFT calculations [53] (Fig. 3).

4.4.1 Create an Input File

In general, to create a Gaussian input file, a text editor like Notepad, TextEdit, or a code editor is opened. The format of this input is quite particular and usually consists of several very important sections. The first section, the Link0 section, is optional and can include things like the amount of memory to be allocated and how many processors the job will take. The critical Route Section defines the type of calculation to be performed (e.g., optimization and frequency analysis), the level of theory to be applied (e.g., the DFT functional and basis set), and other options. This is followed by the Title Section, which is a few lines describing the calculation for the user's later identification and organization.

The core of the input file is constituted by the Molecular Specification section, where the molecular system under investigation is defined, comprising atomic coordinates, the overall charge of the molecule, and its spin multiplicity. Optional sections may be used to constrain some parts of the calculation, provide an initial guess for the calculation, or define other specific options depending on the nature of the calculation. In other words, as you go through this template, inserting the appropriate information with care, you will be writing a well-defined Gaussian input file that describes the calculation at hand [54].

4.4.2 Prepare the Molecular Structure

If you have a 3D structure of the molecule, you can use a molecular modeling software (like Gauss View or Avogadro) to generate the Cartesian coordinates. If you only have a 2D structure, you can use tools like Chem Draw or Marvin Sketch to convert it to 3D coordinates (Fig. 4).

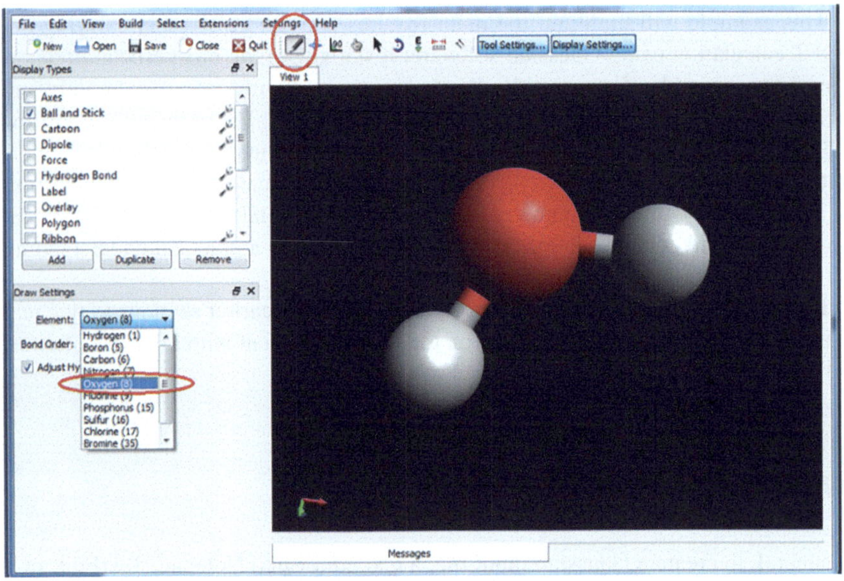

Fig. 4 Avogadro molecular modeling software

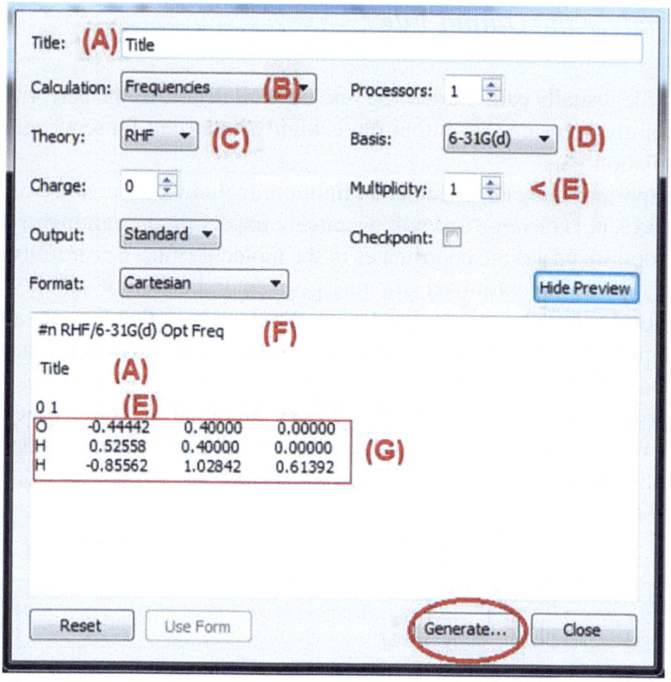

Fig. 5 Interface of Gaussian input

4.4.3 Specify Calculation Parameters

In the Route Section of the input file, define the desired DFT calculation type, functional, and basis set. For example:

#P B3LYP/6-31G(d) opt

This specifies a geometry optimization using the B3LYP functional and 6-31G(d) basis set (Fig. 5).

4.4.4 Submit the Job

Once you have carefully prepared your Gaussian input file, remember to save it with a meaningful name followed by the ".com" suffix, for example, water.com. Next, open a terminal or command prompt, change into the folder where you've saved your input file. Now, simply type in the following, which is a typical command format to run Gaussian: "g09 water.com" This invokes Gaussian to use the mentioned input file, in this case water.com, and execute the calculation specified.

4.4.5 Analyze the Output File

The output file, usually called water.log, includes all details of the performed calculation. Carefully perusing this output file is highly important for securing the result of the calculation.

Some important things to check at a minimum include: convergence of the SCF: if the SCF does not converge this will negatively impact on the validity of the calculation. The optimized atomic coordinates of the molecule must be carefully checked. The total energy of the optimized structure gives an indication of stability. If vibrational frequencies have been calculated it is crucial to check that they are all positive (a negative frequency indicates the structure is a transition state rather than a stable minimum on the potential energy surface).

Other properties that can be derived, depending on the type of calculation performed, will include Mulliken charges, electron density, and molecular orbitals, giving additional insight into the electronic structure and properties of the system under study (Fig. 6).

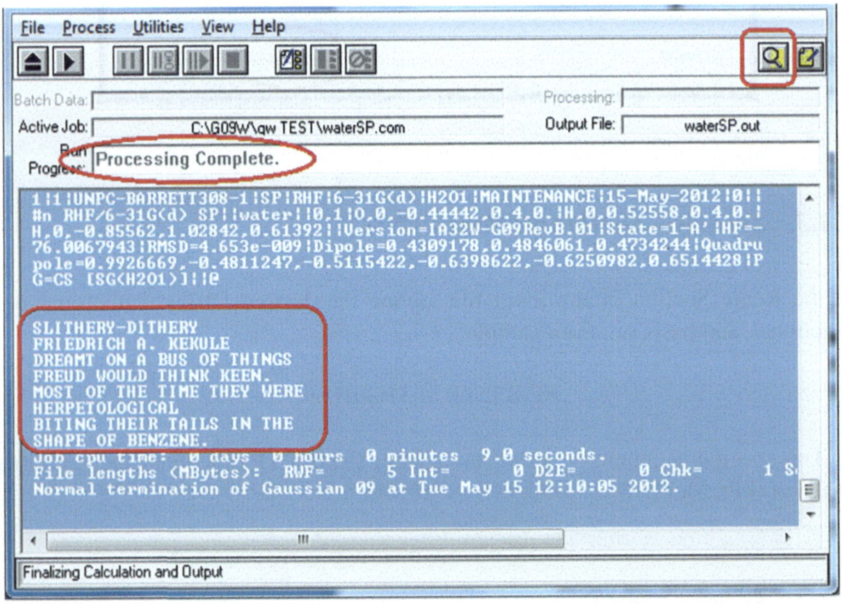

Fig. 6 Interface of Gaussian Output

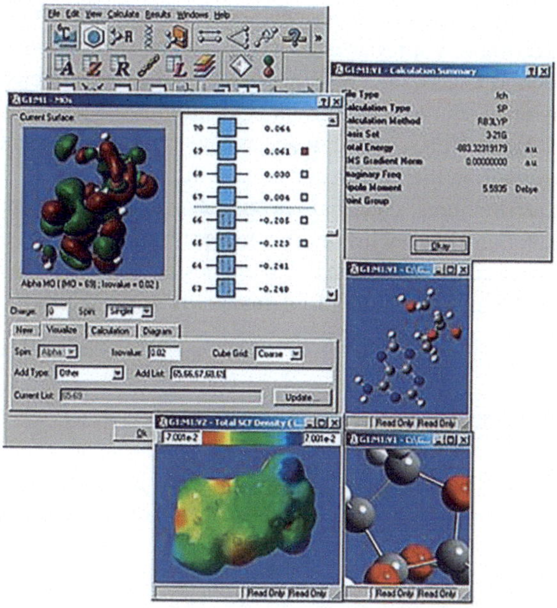

Fig. 7 Gaussian software interface

4.4.6 *Visualize the Results*

Use visualization software like Gauss View or Avogadro to visualize the optimized geometry, molecular orbitals, and other properties (Fig. 7).

Sample Gaussian Input File:

```
#P B3LYP/6-31G(d) opt
 Water Molecule
 0 1
O 0.0 0.0 0.0
H 0.0 0.0 1.0
H 0.0 1.0 0.0
```

Gaussian Software Interface:

Following these steps should allow one to carry out DFT calculations using Gaussian software and then analyze the results obtained for valuable insights into the electronic structure and properties of molecules [55–58].

This chapter has comprehensively illustrated the software tools both commercially available and open-source used for DFT calculations. We have discussed the strengths, weaknesses, and applications of various software packages to provide a clear view for researchers who would seek to tap into the power of DFT.

It has presented in this chapter a well-diversified landscape of DFT tools, from the user-friendly interfaces and very advanced functionalities of commercial software such as Gaussian, VASP, and ADF to the cost-effective and customizable provided by open-source packages like Quantum ESPRESSO, OpenMX, and Yambo.

Key features, applications, and good practice associated with these software packages allow the researcher to take full advantage of DFT in investigating molecular structures, electronic properties, reaction mechanisms, and the design of new materials. In conclusion, there is great potential that the combination of DFT with advanced computational techniques will drive innovation and help solve many challenging scientific problems.

References

1. R.G. Parr, Density functional theory of atoms and molecules BT—Horizons of quantum chemistry. Horiz. Quantum Chem., 5–15 (1980). https://doi.org/10.1007/978-94-009-9027-2_2
2. F. Ju et al., *Acceleration Without Disruption: DFT Software as a Service*, pp. 1–49
3. Physics Research and Technology
4. A.P. Introduction, Density Functional Theory
5. P. Giannozzi et al., *Quantum Espresso: a Modular and Open-Source Software Project for Quantum Simulations of Materials*, pp. 1–18
6. A.A. Ziabari, G.S. Chang, *1-Introduction* (2023). https://doi.org/10.13140/RG.2.2.34092.03203
7. I. Carnimeo et al., Quantum espresso: one further step toward the exascale. (2023). https://doi.org/10.1021/acs.jctc.3c00249
8. M. Olopade, A.B. Adegboyega, A.D. Adewoyin, Nigerian society sciences Investigation of the behaviour of tunable chalcogenide-Bismuth based perovskite BiTl (SxSe $1-x$) $X = 0, 0.33, 0.67, 1$: first. **3**, 0–11 (2025). https://doi.org/10.46481/jnsps.2025.2259
9. S.K. Yadav, Structural, electronic, optical, magnetic, and mechanical properties of $SmMnO_3$ perovskite with europium and yttrium doping: a first-principles study. **125309**, 1–14 (2024). https://doi.org/10.1063/5.0232439
10. K. Kshetri, First Principles Study of Structural, Electronic and Mechanical First Principles Study Of Structural. (2024). https://doi.org/10.13140/RG.2.2.25854.60484
11. A.A. Al-Enzi, Structural, electronic, and mechanical properties of anatase and rutile titanium dioxide phases using the density functional theory. **14**(5), 16860–16865 (2024)
12. C. Mauney, M.B. Nardelli, D. Lazzati, Formation and properties of astrophysical carbonaceous dust. I: ab-initio calculations of the configuration and binding energies of small carbon clusters. (2014)
13. S. Ehsanfar, F. Kanjouri, H. Tashakori, A. Esmailian, First-principles study of structural, electronic, mechanical, thermal, and phonon properties of III-phosphides (BP, AlP, GaP, and InP). J. Electron. Mater. (2017). https://doi.org/10.1007/s11664-017-5640-4
14. J. Miguel, R. Espitia, H. John, F. Díaz, J. Arbey, R. Martínez, Structural and electronic properties of V-doped cubic BN: A density functional theory study. Solid State Commun. **244**, 23–27 (2016). https://doi.org/10.1016/j.ssc.2016.06.014
15. P. Giannozzi, C. Cavazzoni, Large-scale computing with quantum espresso. **32**, 49–52 (2009). https://doi.org/10.1393/ncc/i2009-10368-9
16. S. Boker, M.C. Neale, H. Maes, M. Wilde, OpenMx: an open source extended structural equation modeling framework. (2011). https://doi.org/10.1007/s11336-010-9200-6

17. R.M. Kirkpatrick, R. Estabrook, T.C. Bates, H.H. Maes, HHS public access. **81**(2), 535–549 (2017). https://doi.org/10.1007/s11336-014-9435-8.OpenMx
18. A.N. Labs, A.N. Labs, M n h m. **306–317** (2011)
19. J.N. Pritikin, K.M. Schmidt, Model builder for item factor analysis with OpenMx. **8**, 182–203 (2016)
20. D. Sangalli, A. Ferretti, H. Miranda, C. Attaccalite, I. Marri, Many-body perturbation theory calculations using the yambo code. **31** (2019). https://doi.org/10.1088/1361-648X/ab15d0
21. A. Marini, Optical and electronic properties of copper and silver: from density functional theory to many body effects
22. A. Marini, C. Hogan, M. Grüning, D. Varsano, Yambo: An ab initio tool for excited state calculations ☆. Comput. Phys. Commun. **180**(8), 1392–1403 (2009). https://doi.org/10.1016/j.cpc.2009.02.003
23. E. Molteni, G. Cappellini, D. Sangalli, Ab initio circular dichroism with the Yambo code: applications to dipeptides Ab initio circular dichroism with the Yambo code: applications to dipeptides. https://doi.org/10.1088/1757-899X/1265/1/012005
24. M.V. Malyshkina, A.S. Novikov, *Modern Software for Computer Modeling in Quantum Chemistry and Molecular Dynamics* (2021), pp. 134–144
25. C. Ravi, S. Kumar, Approach of density functional theory to molecules using Gaussian. **15**(1), 1–13 (2019)
26. D.R. Salahub, E. Proynov, *Density Functional Theory Its Gaussian Implementation and Applications to Complex Systems* (2017). https://doi.org/10.1007/978-1-4899-1340-1
27. M. Sharma, M. Sierka, Efficient implementation of density functional theory based embedding for molecular and periodic systems using Gaussian basis functions. (2022). https://doi.org/10.1021/acs.jctc.2c00380
28. M.E. Ayalew, *DFT Studies on Molecular Structure, Thermodynamics Parameters, HOMO-LUMO and Spectral Analysis of Pharmaceuticals Compound Quinoline (Benzo [b] Pyridine)* (2022), pp. 29–42. https://doi.org/10.4236/jbpc.2022.133003
29. A. Ballesteros, *Gaussian Curvature Effects on Graphene Quantum Dots* (2023), pp. 1–15
30. A. Alexander et al., Characterization of surface intermediates in zeolite catalysis the accuracy challenge of the DFT-based molecular assignment of 13 C MAS NMR characterization of surface intermediates in zeolite catalysis †‡. Phys. Chem. Chem. Phys. **22**, 24004–24013 (2020). https://doi.org/10.1039/D0CP04439C
31. J. Hafner, *The Vienna AB-Initio Simulation Program VASP: An Efficient and Versatile Tool for Studying the Structural, Dynamic, and Electronic Properties of Materials* (2015). https://doi.org/10.1007/978-1-4615-5943-6
32. M. Kertesz, G. Kresse, Performance of the Vienna ab initio simulation package (VASP). **624**, 37–45 (2003)
33. G. Kresse, T.U. Wien, *The Vienna Ab-Initio Simulation Program VASP: an Efficient And Versatile Tool For Studying The Structural, Dynamic, And Electronic Properties of Materials* (1997), pp. 69–82
34. V. Eyert et al., Unravelling the potential of density functional theory through integrated computational environments: recent applications of the Vienna Ab initio simulation package in the medea® software. (2018). https://doi.org/10.3390/computation6040063
35. G.T.E. Velde, F.M. Bickelhaupt, E.J. Baerends, C.F. Guerra, S.J.A.V.A.N. Gisbergen, Chemistry with ADF. **22**(9), 931–967 (2001)
36. S.K. Wolff, Amsterdam density functional package. **104**, 645–659 (2005). https://doi.org/10.1002/qua.20653
37. ADF Property Programs. pp. 1–38 (2009)
38. C.C. Pye, T. Ziegler, Regular article an implementation of the conductor-like screening model of solvation within the Amsterdam density functional package, 396–408 (1999). https://doi.org/10.1007/s00214990m184
39. T. Chemistry, *Tutorial on ADF* (2016)
40. A. Tomberg, G 09w t
41. 43.ExploringChemistryWithElectronicStructureMethods.pdf

42. J.W. Andzelm, Density Functional Methods in Chemistry
43. On leave from Facultad de Quimica, UNAM, M´ exico. On leave from Institute of Kinetics and Catalysis, 1113 Sofia, Bulgaria 1. pp. 1–53
44. J. Vandevondele, J. Hutter, Gaussian basis sets for accurate calculations on molecular systems in gas and condensed phases. **114105** (2007). https://doi.org/10.1063/1.2770708
45. J.A.N. Andzelm, J.A.N. Andzelm, Optimization of Gaussian-type basis sets for local spin density functional calculations. Part I. Boron through neon, optimization technique and validation. (19)
46. C.E. Caballero-agosto, E. Caballero, Optimizing a Molecule in GaussView and Gaussian
47. S.C. North, K.R. Jorgensen, J. Pricetolstoy, A.K. Wilson, Population analysis and the effects of Gaussian basis set quality and quantum mechanical approach: main group through heavy element species. (2023). https://doi.org/10.3389/fchem.2023.1152500
48. P. Gunawardhana, Y. Balasooriya, M.S. Kandanapitiye, Y.C. Chau, Applied sciences optoelectronic characterization of natural dyes in the quest for enhanced performance in dye-sensitized solar cells: a density functional theory study. (2024)
49. P. Gao, X. Yang, Y. Tang, A. Anderson, Graphical Gaussian process regression model for aqueous solvation free energy prediction of organic molecules in graphical Gaussian process regression model for aqueous solvation free energy prediction of organic molecules in redox flow battery †. (2021). 10.1039/cXCP00000x/ter
50. M. Krone, J.E. Stone, T. Ertl, K. Schulten, Fast visualization of Gaussian density surfaces for molecular dynamics and fast visualization of Gaussian density surfaces for molecular dynamics and particle system trajectories. (2012). https://doi.org/10.2312/PE/EuroVisShort/EuroVisShort2012/067-071
51. H. Kaneko, Estimation and visualization of process states using latent, 326–333 (2021). https://doi.org/10.1002/ansa.202000122
52. K.N. Kudin, B. Johnson, Linear scaling density functional theory with gaussian orbitals and periodic boundary conditions by requirements for the degree linear scaling density functional theory with gaussian orbitals and periodic boundary conditions by
53. M. Bursch, J. Mewes, A. Hansen, S. Grimme, Best-practice DFT protocols for basic molecular computational chemistry ** angewandte. **202205735** (2022). https://doi.org/10.1002/ange.202205735
54. S.I.D. Presley, V. Babu, A.R. Veeraraghavapuram, T. Nadu, Research on Chemical Intermediates Spectral, Density functional theory calculations, molecular docking studies, antimicrobial, antioxidant, antidiabetic activities of the novel (E)-2-(3-(5-(3-nitrophenyl)furan-2-yl)acryloyl)-3,4-dihydronaphthalen-1(2H)-one. 1–37 (2023)
55. Quantum chemistry
56. M. Towler, (1.1) is the electronic Hamiltonian. For systems of interest in chemistry, analytic where. 1–40 (2000)
57. C. Skylaris, CHEM6085: density functional theory lecture 8 Gaussian basis sets solving the Kohn-Sham equations on a computer
58. R. Geetha, S. Nair, A. Kumar, N. Nair, S. Sun, Density functional theory study of doped coronene and circumcoronene as anode materials in lithium-ion batteries. Sci. Rep. **1–17** (2024). https://doi.org/10.1038/s41598-024-66099-6

Chapter 5
Interpreting DFT Results

Deshani Perera and Roshan Thotagamuge

Abstract Density functional theory (DFT) has become an indispensable computational tool for elucidating the electronic structure and physicochemical properties of molecules and materials. This chapter provides a comprehensive guide to interpreting DFT results with an emphasis on connecting computational outputs to chemical reactivity, stability, and functional performance. Key concepts such as frontier molecular orbitals (HOMO–LUMO), band gap analysis, and global reactivity descriptors are discussed as fundamental indicators of molecular behavior. The chapter further explores advanced electronic structure analyses including density of states (DOS), partial DOS, and band structure calculations to understand charge transport and electronic transitions. Visualization tools such as electrostatic potential maps, electron localization functions, noncovalent interaction analysis, and reduced density gradient plots are presented to reveal intermolecular interactions and bonding characteristics. Applications of these interpretations in materials design, sensing, photocatalysis, and drug delivery systems are highlighted, demonstrating how DFT serves as a powerful bridge between theoretical predictions and experimental observations.

Keywords Noncovalent interaction analysis · HOMO–LUMO analysis · Electrostatic potential maps · Reduced density gradient plots · Molecular reactivity · Electron localization functions

D. Perera (✉)
Department of Nano Science Technology, Faculty of Technology, Wayamba University of Sri Lanka, Kuliyapitiya, Sri Lanka
e-mail: mideshaniperera99@gmail.com

R. Thotagamuge (✉)
Department of Optometry, Faculty of Allied Health Sciences, University of Sri Jayewardenepura, Nugegoda, Sri Lanka
e-mail: roshan@sjp.ac.lk

R. Thotagamuge (ed.), *Density Functional Theory*,
https://doi.org/10.1007/978-981-95-9745-1_5

5.1 Introduction

The density functional theory (DFT) has become a tool in computational chemistry that numerous researchers employ to determine the electronic arrangement of a molecule or material. Discussing the obtained DFT results, this chapter pays more attention to the frontier molecular orbitals (FMOs) which are the highest occupied molecular orbital and the lowest unoccupied molecular orbital. The theory which is referred to as the HOMO–LUMO theory was first put forward in 1952 by a chemist called Kenichi Fukui, and it is relied upon when predicting a chemical's reactivity, stability of the molecule, or even when predicting an electronic transition of the molecule. As referenced above, the energy between the highest occupied and the lowest unoccupied orbital is called the bandgap and it is used to determine how stable and how reactive a compound can be. The HOMO and the LUMO energies can also determine the molecular reactivities of compounds, and these include the ionizing energies, the electronegativity, and the electron affinity (EA). Other topics captured by the chapter include the density of states (DOS) and the studies of band structure these two are important as they explain the electronic structures of the materials while explaining the band gaps and the electronic transitions in the doped regions. Succeeding concepts broaden the knowledge of charge distribution and are likely to result in reactions including ESP maps and electron localization functions which directly visualize the electrostatic potential (ESP) energy.

5.2 Analyzing Electronic Structure

5.2.1 HOMO LUMO and Bandgap (E_g)

FMOs are crucial for predicting and interpreting chemical phenomena such as reaction mechanisms, molecular stability, and electronic transitions which were established by Kenichi Fukui in 1952 [1] and also applied in quantum studies of molecules to investigate electronic level structure, optical, and electrical properties [2, 3]. The terms E_{HOMO} and E_{LUMO} refer to the energy of the highest occupied molecular orbitals (HOMO) and lowest unoccupied molecular orbitals (LUMO), respectively [4].

In the study of HOMO and LUMO, energy levels of molecule advise a charge distribution probability as well as charge transfer in it. The HOMO is the energy level at which electrons are usually available and acts as an electron donor, whereas the LUMO represents levels that can take in electrons which indicates levels associated with electron transitions [5]. As shown in Fig. 5.1, LUMO takes electrons at C = C within the ring and facilitates the electronic transfer from one region of the ring bond to other regions of the pyrimidine bonds, HOMO's electron donor is restricted to C–H, C–C, and C–N bonds [6].

Difference in energy level between HOMO and LUMO represents an energy gap which evaluates reactivity and stability generally with the lowest electronic

(a)

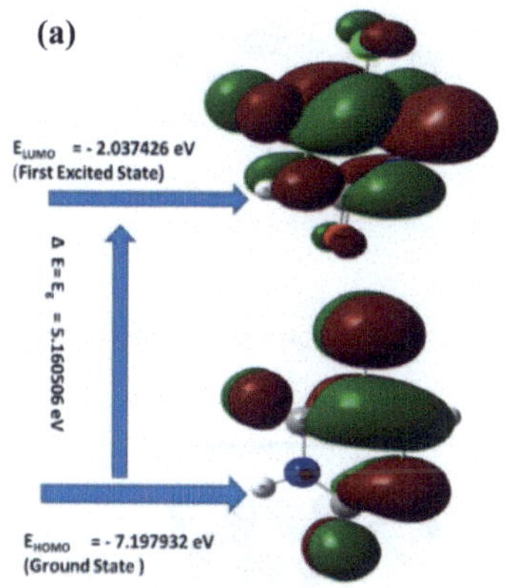

(b)

Parameter	Energy (eV)
HOMO (for the ground level)	– 7.197932 eV
LUMO (for the first excited level)	– 2.037426 eV
HOMO–LUMO = the energy gap(Δ E)	**– 5.1605067 eV**

Fig. 5.1 **a** Electronic energy levels with frontier MOs for 5-chlorouracil **b** Lowest level energy value of HOMO and LUMO of 5-chlorouracil

energy level excitation of biomolecule [7]. A molecule with a large frontier orbital gap (HOMO–LUMO energy gap) has poor chemical reactivity and high kinetic stability because the process of transferring an electron to the high-lying LUMO is energetically unfavorable [8]. Band gap can be calculated by using the following equation:

$$E_g = E_{LUMO} - E_{HOMO} \tag{5.1}$$

According to Fig. 5.1b, relatively small band gap (-5.1605067 eV) suggests that 5-fluorouracil can easily facilitate charge transfer, enhancing its reactivity and stability.

Pengfei Wang et al. studies suggest that, as shown in Fig. 5.2 gives the graph with the ascertained energy gap values, for sensors 1 and 1–Ga^{3+} respectively. The energy gap reduces from about 3.91 eV for the sensor to about 3.36 eV for the complex. This reduces the energy gap; hence, it implies that the tendency of the electrical transitions takes place more readily (less energy is required) when there is Ga^{3+} around. Thus, there was a red-shift in the absorption spectrum from 1 to 1–Ga^{3+}. The band gap (E_{gap}) becomes narrower from 1 to 1–Ga^{3+}, meaning more and more the maximum wavelength of the absorption will become broader and is in accordance with the experimental results [9].

The electron density isosurfaces of the HOMO and LUMO are shown in Fig. 5.3 for the FU@C_6N_6 and NU@C_6N_6 complexes, which are significant in determining various properties of the compounds. The red isosurfaces appear on the negative part of the wavefunctions which means electrons are distributed more in these regions, and the green isosurfaces appear on the positive part of the wavefunctions where the electrons are distributed less. In the FU@C_6N_6 model, the charge density in the HOMO orbital is mainly generated by the FU molecule, meaning that there exists an effective overlap of electronic interactions while the LUMO is only centered on C_6N_6, a typical electron acceptor type molecule. From the NU@C_6N_6 model, the

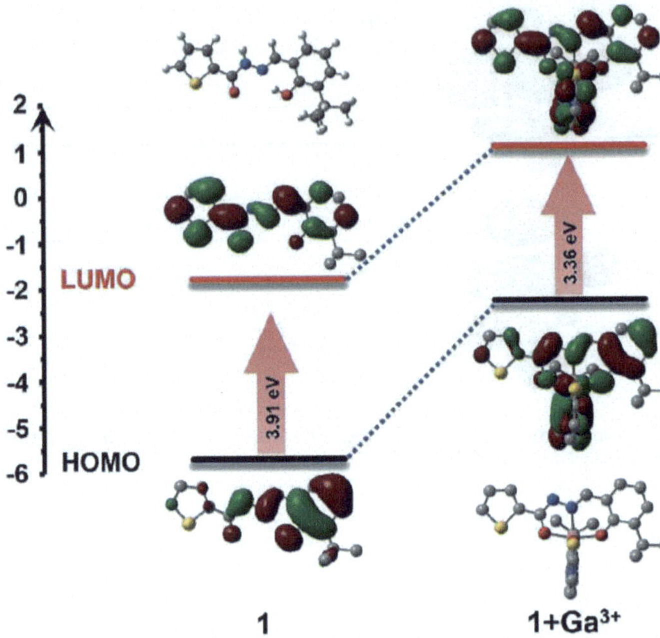

Fig. 5.2 Optimized molecular structure and HOMO and LUMO orbital levels and energy gap of 1 and 1 + Ga^{3+}, calculated with DFT at the B3LYP/6-31G(d) level using Gaussian 09

HOMO is, however, centered toward the NU molecules while the LUMO still rests on C_6N_6. This distribution further emphasizes the charge transfer mechanism of the drugs and the concave C_6N_6 particle, where a large reduction in the HOMO–LUMO energy gap is caused in the FU@C_6N_6 complex, which is good for the drug delivery system since it must be conductive and reactive [10].

5.2.2 *Chemical Reactivity Parameters*

The global reactivity parameters are defined within the context of the energies of the frontier molecular orbitals (EH and EL) and are given by the following relations [11]. In the process of interaction with molecules, LUMO accepts electrons and the corresponding energy is known as the EA while the appearing HOMO denotes the electron donors, whose energy corresponds to ionization potential (IP) [12]. If the energy gap of molecule is small, then it has good chemical reactivity, high polarizable, and softness. The Koopman theory was used to determine the physicochemical parameters of each configuration by defining the energies of the LUMO and HOMO using the following equations for chemical potential (μ), chemical hardness (η), chemical softness (S), electronegativity (χ), and electrophilic index (ω) from the

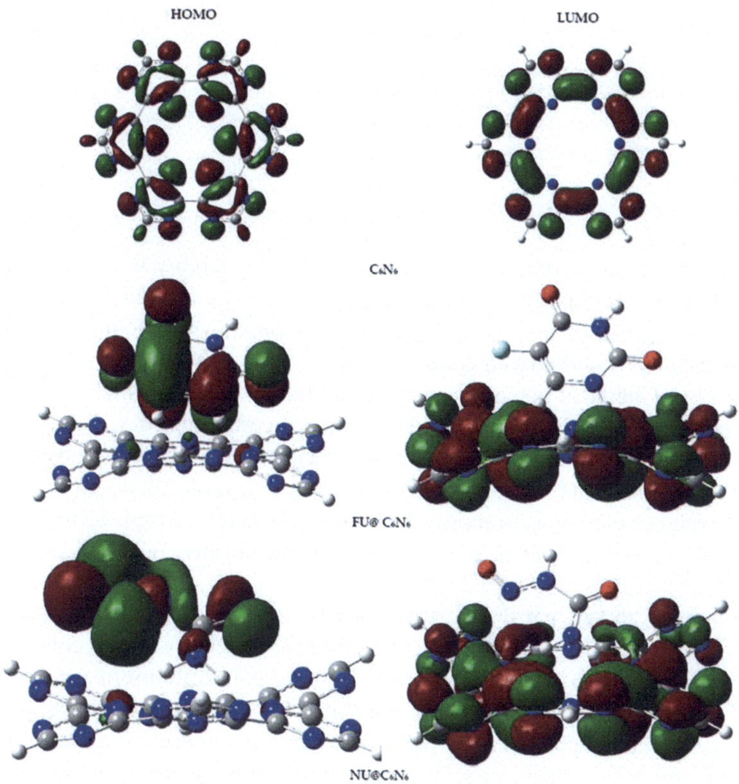

Fig. 5.3 Orbital densities of FU@C6N6 and NU@C6N6 complexes obtained through FMO analysis (isovalue = 0.02 a.u.)

following equations [13, 14]. ΔN_{max} indicates the maximum amount of electronic charge that can be absorbed by a system [15].

$$IP = -E_{HOMO} \tag{5.2}$$

$$EA = -E_{LUMO} \tag{5.3}$$

$$\mu = (E_{HOMO} + E_{LUMO})/2 \tag{5.4}$$

$$\eta = (E_{LUMO} - E_{HOMO})/2 \tag{5.5}$$

$$\omega = \mu^2/2\eta \tag{5.6}$$

$$S = 1/\eta \tag{5.7}$$

$$\Delta N_{max} = -\frac{\mu}{\eta} \tag{5.8}$$

$$\chi = -\mu = -(E_{HOMO} + E_{LUMO})/2 \tag{5.9}$$

The electronegativity index (ω) measures a molecule's ability to attract electrons (nucleophilic or electrophilic character) [16]. The negative chemical potential reflects on the stability of a compound. ΔN_{max} indicates maximum amount of electronic charge that can be absorbed by a system. A positive ΔN_{max} means that it is an electron acceptor [17]. Chemical potential is a thermodynamic variable which is defined by energy change during the transfer of electrons in a certain reaction [18]. Chemical potential (μ) is a measure of the propensity of electrons to leave the given system [19]. On the other hand, the tendency of an atom or a molecule to pull toward its nuclei its valence electrons, or more generally their shared valence electrons termed as electronegativity (χ), can also be expressed as the negative of chemical potential (μ) [20].

Additionally, the greater the electronegativity of an atom, the more reactive it is also expected to be. Another variable affecting the reactivity of the compound is the chemical hardness (η), which characterizes the polarizability of a molecule or the ability for charge transfer. Broadly speaking, this means that the lower η values means that the electrophile is more polarizable [21]. A molecule's softness (S/σ) is determined by its polarizabilities. It should be mentioned that soft molecules with narrow HOMO–LUMO energy gaps are effective corrosion inhibitors because they can readily donate electrons to metals [22]. Along with a large positive value of μ, which denotes the electrophilic character of the compounds mentioned and in contrast to polarized on low negative value of ω signifying nucleophilic behavior of the molecules [23].

As per the Ajaz Hussain et al. studies, compounds 1, 2, and 3 refer to three novel phenyl pyrimidine derivatives that were synthesized and studied for their structural, electronic properties. According to Table 5.1 data, the affinity for electrons of compound 1 is the highest (2.818 eV) coupled with global softness (0.486), this means it is more reactive and has an increased capability of accepting electrons than these other compounds. Conversely, compound 3 has shown the highest IP of 6.002 eV and global hardness, 1.962 eV, meaning it is of the highest stability and the least in reactivity among the three compounds. Further on, compound 3 is also the highest in electronegativity (4.040 eV) suggesting better electron drawing ability. In summary, it can be noted that compound 1 can actively take part in charge transfer and is therefore more reactive whereas compound 3 can remain without toxic effects because it is less reactive and stable in nature [24].

Table 5.1 Ionization potential (I), electron affinity (A), electronegativity (X), global hardness (η), chemical potential (μ), global electrophilicity (ω), and global softness (σ) values in eV for compounds **1**, **2**, and **3**

Derivatives	I	A	χ	η	μ	ω	σ/S
1	4.874	2.818	3.846	1.028	−3.846	7.174	0.486
2	4.627	2.521	3.574	1.053	−3.574	6.065	0.474
3	6.002	2.078	4.040	1.962	−4.040	4.159	0.254

5.3 Density of States (DOS)

The energy DOS of a molecule shows the number of electron energy states that can be occupied in the molecule. Whole orbital systems can be described with the aid of such plots [25]. In the next DOS plots, for example, the energy scale is plotted along the x-axis, while the y-axis displays the density of states or the number of electronic states at each energy level [26]. The two dominant atoms that contain curcumin's HOMO orbital are the double bond atoms (–C=C–) of the aryl rings, the vinyl groups, and the oxygen in the methoxy and hydroxyl groups. When curcumin is attached to CNT (8,0–6) nanotube, the energy gap decreases from 3.58 eV (curcumin) to 3.02 eV (after interaction). This shift in energy gap is an indication of the strength of the electron interactions after the curcumin is adsorbed onto the nanotube which suggests it will increase the reactivity and the electronic properties [27] (Fig. 5.4).

Partial Density of State (PDOS)

The TDOS describes all the electronic states in the system while the PDOS differentiates these states by atomic levels enabling one to find the contributing atoms or orbitals in the electronic structure at the different energy states [28]. In Fig. 5.5, it can be seen that there is a new band of energy level which has been doped in the structure this means that the dopant bismuth oxide (BiO_2-X) interacts with $AgVO_3$ such that the band of the material attaches itself and electrons are transferred between states [29]. With the increase in the PDOS on the edges of the conduction band, more excitations of the electrons are possible which results in the enhancement of carrier concentration and electronic conductivity leading to better photocatalytic activity of the heterostructure [30].

5.4 Band Structure (BS)

Important details on the energy range that electrons can occupy (energy bands) or the energy areas that they cannot occupy (band gap) are provided by electronic band structure. The electronic band gap in insulators and semiconductors is measured as the difference and possible transitions between the valence band maximum (VBM) and the conduction band minimum (CBM) [31].

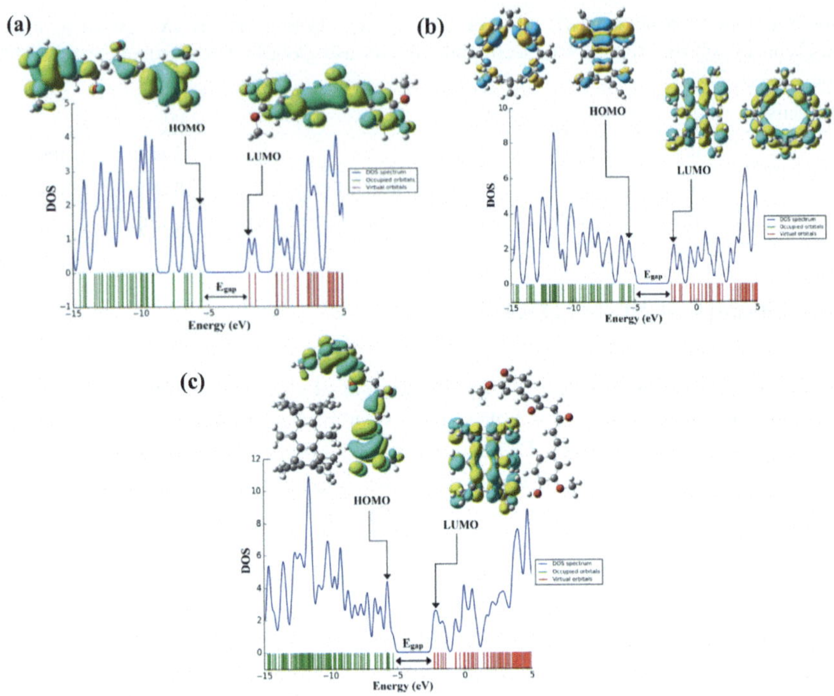

Fig. 5.4 Calculated HOMO, LUMO orbitals, and DOS plots of the **a** compound curcumin **b** CNT (8,0–6) **c** CNT (8,0–6)/curcumin

This intrigued with the band structure of both pure SrTi O_3 and Cu-doped SrTi O_3 to see how doping affects the electromagnetic properties of the material. Phonon assistance in electron excitation transitions is shown in Fig. 5.6a where the pure SrTi O_3 BS shows an indirect band gap also known as CBM, located at the G point and the VBM at the R point. It creates energy losses and lower efficiency that results in a band gap of around 1.812 eV which is lower than the reported experimental value of 3.2 eV from a generalized gradient approximation (GGA) as its cause. It is different in the case of Cu-doped $SrTiO_3$ which features a linear configuration where both the CBM and the VBM are positioned at the same G point therefore transforming the material into a direct band gap semiconductor. The modification in copper substitutional d states promotes the efficiency of the electron excitation transition which further improves the photocatalytic performance and the optical properties of the substrate [32, 33].

However, the band structure analysis is essential for elucidating the electron characteristics of the material and its potential use in spintronics applications [34]. Figure 5.7 presents the band structure of $LaCoO_3$, revealing distinct electronic properties in spin-up (SU) and spin-down (SD) states. The SU state features an indirect band gap of 3.0 eV, indicating that the CBM and VBM occur at different momentum

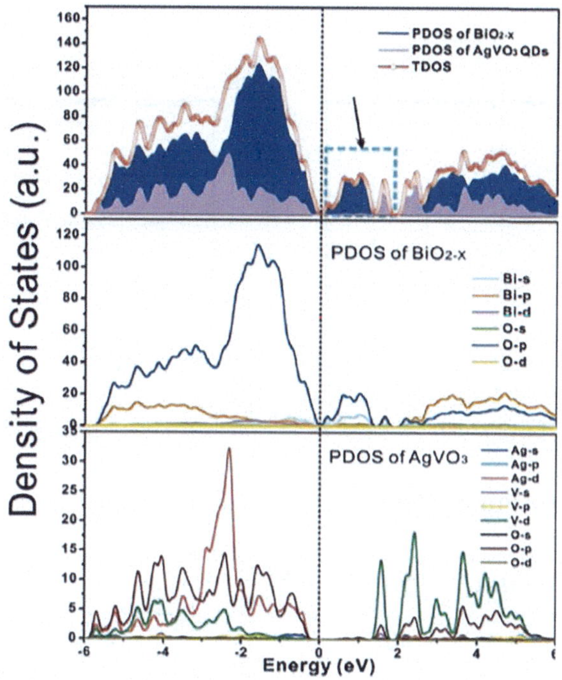

Fig. 5.5 Calculated TDOS and the PDOS of $AgVO_3/BiO_2$-X heterostructure utilizing the DFT calculations method

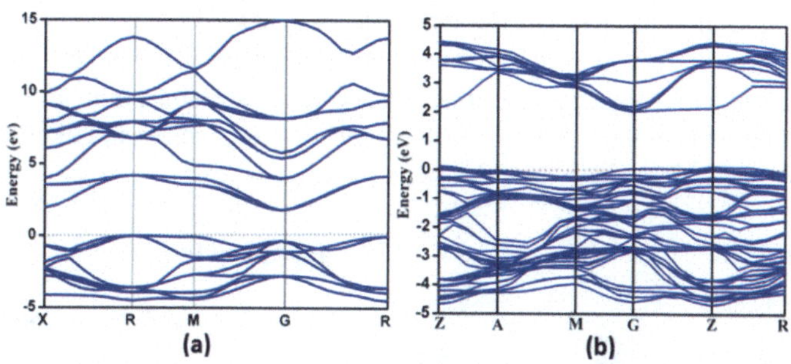

Fig. 5.6 **a** Band Structure of pure $SrTiO_3$ and **b** Cu-doped $SrTiO_3$

values, which suggests insulating behavior. In contrast, the SD state has a direct band gap of 1.0 eV, implying potential metallic characteristics [35].

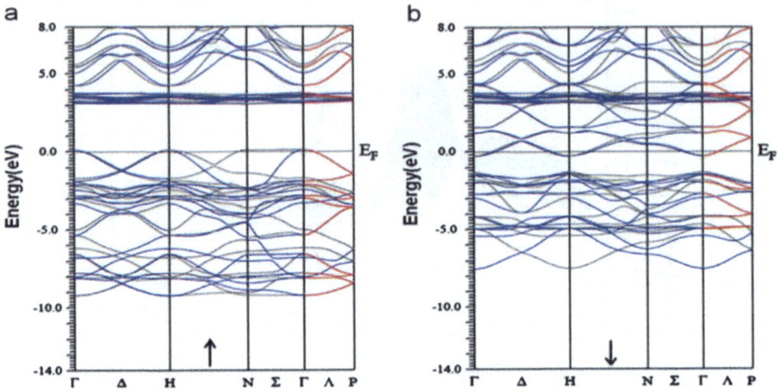

Fig. 5.7 Band structure for $LaCoO_3$ in **a** SU and **b** SD states

5.5 Visualization of Results

5.5.1 *Electrostatic Potential—ESP*

Ataei et al. studies show how chitosan nanoparticles are used to deliver favipiravir using ESP maps which are used to evaluate the distribution of charge in molecules and identify potential reaction sites as shown in Fig. 5.8. According to the ESP map for favipiravir (FP), this region has experienced one peak around the NH_2 group showing that it is capable of nucleophilic attack although there are other peaks but higher than that, in other parts of the molecule as well [36].

Furthermore, there was a global minimum point near OH attached to the ring structure showing that electrophilic attack can be targeted at this point. These findings suggest certain parts of the FP molecule that may be more reactive and likely to engage

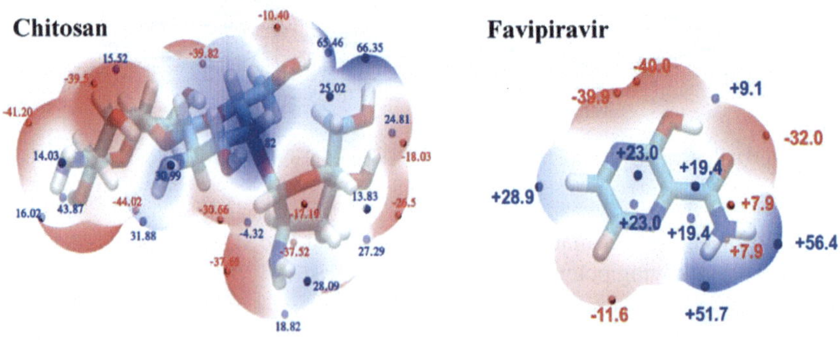

Fig. 5.8 The optimized chitosan nanoparticles and favipiravir molecule electrostatic potential energy surface (ESP)

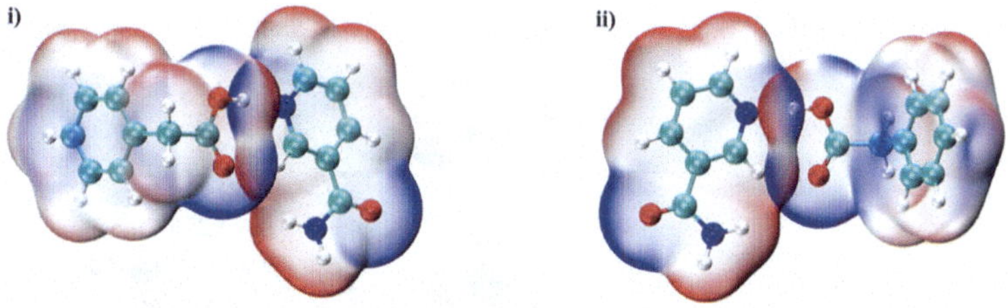

Fig. 5.9 EPS of PN1: (i) front view and (ii) opposite view. The blue areas represent negative charges, while the red areas are positively

in chemical interactions [37]. The red points stand for local minima, while the blue ones signal local maxima points on the ESP surface.

Drawing evidence from Kusuma et al., in the case of tubular cocrystals as well as PYCNIC cocrystals intermolecular hydrogen bonds made of electronegative atoms are observed to be common and most tend to oppose each other (Fig. 5.9).

From the discussions presented in Fig. 5.8, the hydrogen atom attached to one atom acts as a donor and is positively charged while the oxygen atom in the water molecule formed is clearly the hydrogen bond acceptor; therefore, hydrogen possesses a negative charge. The same case is also seen in any CPHS synthon where the O–H unit from PYC which is positively charged interacts with the N atom from NIC which is negatively charged. The same goes for weaker C–H–O bonds where O bearing the negative charge will bind with C–H from PYC [38].

5.5.2 Noncovalent Interactions—NCI

There is a significant impact of noncovalent interactions (NCI) between drug and carrier on drug release at the target site. NCI analysis is very important to identify the intermolecular interactions based on the electron densities of the complexes [25]. The NCI analysis comprised the two-dimensional reduced density gradient (RDG) graphs and three-dimensional isosurfaces of the complexes [6]. To identify specific areas, two functions sign ($\lambda 2$) (r)p(r) and RDG can be used. If sign ($\lambda 2$) $\rho < 0$, it means there is an attractive interaction like hydrogen bonding. If sign ($\lambda 2$) $\rho > 0$, it implies a repulsive interaction like steric repulsion.

Finally, if the sign ($\lambda 2$)$\rho \approx 0$, it indicates the presence of van der Waals (vdW) interactions which arise from overlapping electron clouds and occur over larger distances [26]. According to the Medimagh et al. studies, it is important to maintain dispersion as it ensures uniform distribution of drug molecules and prevents molecules from aggregating within the carrier matrix thereby steric repulsions support controlled drug release from the delivery system. Figure 5.10 shows the van der Waal, steric repulsions, and hydrogen bonding in green, red, and blue colors, respectively [39].

Hydrogen bonding between drug molecules and carrier matrix can either slow down or speed up the release of drugs from this system. Strong hydrogen bonds

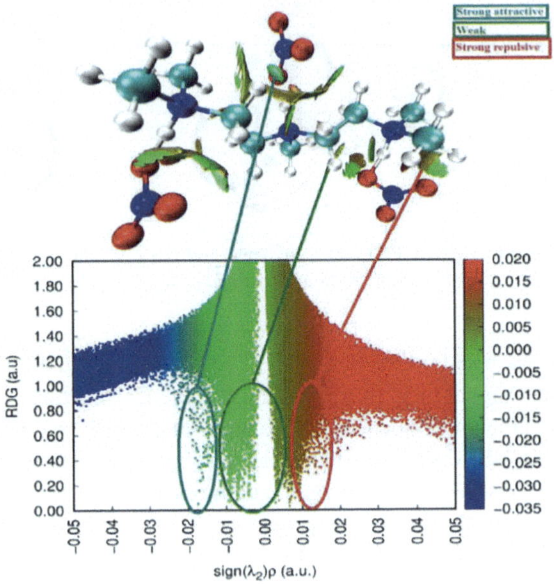

Fig. 5.10 RDG images of $C_9H_{26}N_3\ (NO_3)_3$

may retain drugs efficiently within a given system hence delaying their release while weaker interactions may promote dissociation and hence probably enhance the release [40].

5.5.3 Electron Localization Function—ELF

ELF density also known as the electron localization function that is related to kinetic energy density and correlated with Pauli repulsive force provides a clear picture of electrons distribution in bonds [41]. The examination of the density gradient field allows for the identification of attractive regions and their corresponding basins. The ELF and LOL analyses, which rely on kinetic energy density are carried out using Multiwfn software [42].

From Fig. 5.11, it can be concluded that the molecule exhibits significant electron localization around hydrogen atoms, indicating strong covalent bonding. The delocalized electron clouds around carbon atoms suggest a stable molecular structure with potential resonance [43].

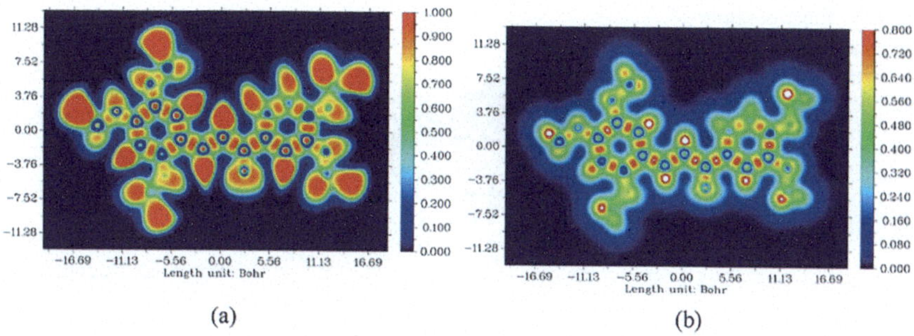

Fig. 5.11 ELF and LOL of molecule presented in color shade map

5.5.4 Ramachandran Plots

Ramachandran plot provides a simple two-dimensional graphic representation of all possible protein structures in terms of torsion angles [44]. Figure 5.12 indicate the distribution of amino acid residues in the allowed and disallowed regions of the phi-psi torsional angles [45]. A significant majority (90%) of residues are located in the allowed regions, suggesting good structural stability and favorable conformations for all proteins. This stability is crucial for effective binding interactions with the title compound, enhancing its potential bioactivity [46].

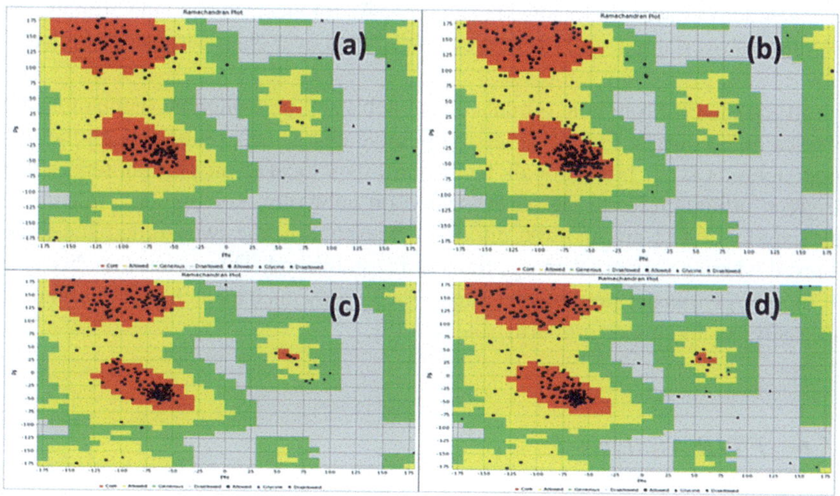

Fig. 5.12 Ramachandran plot for **a** antiviral protein, **b** antibacterial protein, **c** anticanceral protein **d** antimalarial protein

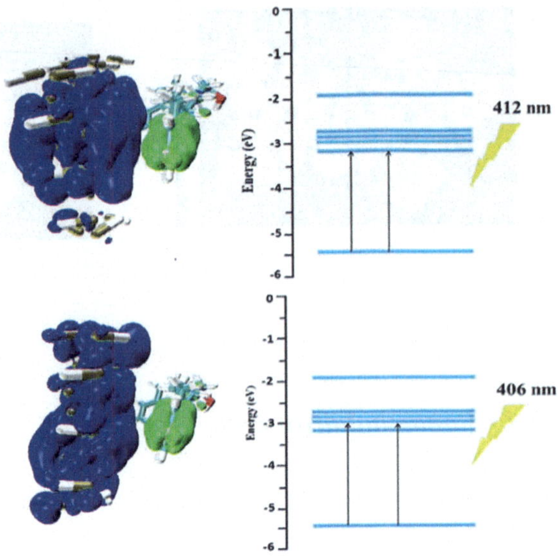

Fig. 5.13 Electron–hole orbitals and PET analysis of two states of nebivolol-PH complex

5.5.5 *Photoinduced Electron Transfer (PET) Analysis*

As per the PET analysis, the greater distance (D) between the hole and electron orbitals will lead to an increase in the length of the charge transfer. The distance (S) is localized lays lower for localized excitation. As for electron and hole orbitals, the low value of overlap integral (S) of the two depicted over charge transfer's high rate [47].

Further, the electron–hole orbitals of the nebivolol-phosphorene (nebivolol-PH) complex are shown in Fig. 5.13 where effective charge separation around the two components is seen; blue orbitals being the holes located on phosphorene while the green orbitals refer to electrons located on nebivolol. For the PET, maximum wavelengths of 412 nm and 406 nm that correspond to the two excited states are achieved where a very good charge transfer is established. This means that, because it is the charge transfer process that affects the fluorescence quenching, the complex will easily be able to perform drug release [48].

5.5.6 *FTIR Analysis*

Infrared spectroscopy probes the molecular vibrations. Functional groups can be associated with characteristic infrared absorption bands, which correspond to the fundamental vibrations of the functional groups [49]. Yu et al. studies demonstrate the HA (hyaluronic acid)-based nano-carrier containing doxorubicin (DOX) and cisplatin (CDDP) as a CD44-targeting anti-cancer drug delivery system, followed

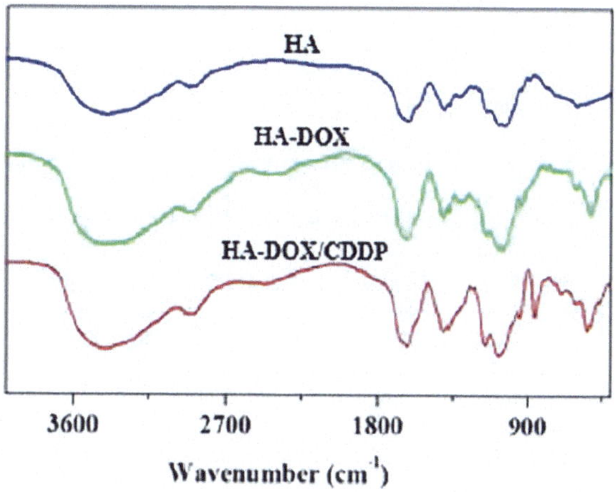

Fig. 5.14 Fourier transform IR spectra of HA, HA-DOX, and HA-DOX-CDDP

by studying its tumor. HA-DOX, HA-DOX-CDDP, and HA can be seen in Fig. 5.14 FTIR spectra. The absorption bands of FTIR spectra for HA-DOX and HA-DOX-CDDP were due to the C-O bond stretching vibrations of DOX which fell in the region of 1300 –1000 cm^{-1} [50]. However, the spectrum for HA-DOX-CDDP also showed a distinctive absorption band at around 3300 to 3200 cm^{-1} due to primary amine (NH_2) groups [51].

5.6 Conclusion

In conclusion, it has been established that DFT interpretation is important in the analysis of compound electronic structure and properties. It has been shown that by studying frontier molecular orbitals, energy gaps, and global reactivity parameters, it is possible to model the chemical behavior of the material and improve it for wide applications. Other quantitative techniques such as density of states, band structure, and ESP maps help understand molecular interactions better, which assists even more in designing efficient drug delivery systems and improving the properties of materials in a myriad of scientific applications.

References

1. A. Sengupta, B. Li, D. Svatunek, F. Liu, K.N. Houk, Cycloaddition Reactivities analyzed by energy decomposition analyses and the frontier molecular orbital model. Acc. Chem. Res. **55**(17), 2467–2479 (2022). https://doi.org/10.1021/acs.accounts.2c00343

2. P. Gunawardhana, Y. Balasooriya, M. S. Kandanapitiye, Y. F. C. Chau, M. R. R. Kooh, R. Thotagamuge, Optoelectronic characterization of natural dyes in the quest for enhanced performance in dye-sensitized solar cells: a density functional theory study. Appl. Sci., **14**(1), (2024). https://doi.org/10.3390/app14010188
3. H. Pir, N. Günay, D. Avci, and Y. Atalay, Molecular structure, vibrational spectra, NLO and NBO analysis of bis(8-oxy-1-methylquinolinium) hydroiodide. Spectrochim. Acta - Part A Mol. Biomol. Spectrosc., **96**, pp. 916–924, (2012) https://doi.org/10.1016/j.saa.2012.07.044
4. L.M.J. Sadeq et al., Examination of the adsorption mechanism of procarbazine anti-cancer drug onto the functionalized GaN nano-tube as a drug delivery vehicle: a DFT study. Bull. Mater. Sci. **46**(3), 1–7 (2023). https://doi.org/10.1007/s12034-023-02979-3
5. M. Raftani, T. Abram, N. Bennani, M. Bouachrine, Theoretical study of new conjugated compounds with a low bandgap for bulk heterojunction solar cells: DFT and TD-DFT study. Results Chem. **2**, 100040 (2020). https://doi.org/10.1016/j.rechem.2020.100040
6. J.S. Singh, M.S. Khan, S. Uddin, A DFT study of vibrational spectra of 5-chlorouracil with molecular structure, HOMO–LUMO, MEPs/ESPs and thermodynamic properties. Polym. Bull. **80**(3), 3055–3083 (2023). https://doi.org/10.1007/s00289-022-04181-7
7. M.E. Ayalew, DFT Studies on molecular structure, thermodynamics parameters, HOMO-LUMO and spectral analysis of pharmaceuticals compound quinoline (Benzo[b]Pyridine). J. Biophys. Chem. **13**(03), 29–42 (2022). https://doi.org/10.4236/jbpc.2022.133003
8. M. Miar, A. Shiroudi, K. Pourshamsian, A.R. Oliaey, F. Hatamjafari, Theoretical investigations on the HOMO–LUMO gap and global reactivity descriptor studies, natural bond orbital, and nucleus-independent chemical shifts analyses of 3-phenylbenzo[d]thiazole-2(3H)-imine and its para-substituted derivatives: Solvent and subs. J. Chem. Res. **45**(1–2), 147–158 (2021). https://doi.org/10.1177/1747519820932091
9. P. Wang, F. Meng, H. Su, L. Liu, M.A. Khan, H. Li, A highly selective 'turn-on' water-soluble fluorescent sensor for gallium ion detection. RSC Adv. **11**(32), 19747–19754 (2021). https://doi.org/10.1039/d1ra02582a
10. M. A. Alkhalifah, M. Yar, I. Bayach, N. S. Sheikh, K. Ayub, Covalent organic framework (C6N6) as a drug delivery platform for fluorouracil to treat cancerous cells: a DFT study. Materials (Basel). **15**(21). (2022) https://doi.org/10.3390/ma15217425
11. R. Pal, P.K. Chattaraj, Chemical reactivity from a conceptual density functional theory perspective. J. Indian Chem. Soc. **98**(1), 100008 (2021). https://doi.org/10.1016/j.jics.2021.100008
12. K. Fukui, T. Yonezawa, H. Shingu, A molecular orbital theory of reactivity in aromatic hydrocarbons. J. Chem. Phys. **20**(4), 722–725 (1952). https://doi.org/10.1063/1.1700523
13. Z. Jafari, R. Baharfar, A.S. Rad, S. Asghari, Potential of graphene oxide as a drug delivery system for Sumatriptan: a detailed density functional theory study. J. Biomol. Struct. Dyn. **39**(5), 1611–1620 (2021). https://doi.org/10.1080/07391102.2020.1736161
14. E. Barim, F. Akman, Synthesis, characterization and spectroscopic investigation of N-(2-acetylbenzofuran-3-yl)acrylamide monomer: Molecular structure, HOMO–LUMO study, TD-DFT and MEP analysis. J. Mol. Struct. **1195**, 506–513 (2019). https://doi.org/10.1016/j.molstruc.2019.06.015
15. S. A. Jasim et al., Nanostructures of boron nitride: A promising nanocarrier for anti-cancer drug delivery. Micro Nanostructures, **185**(2023), p. 207708, (2024), https://doi.org/10.1016/j.micrna.2023.207708
16. M. Shamim et al., DFT study of therapeutic potential of graphitic carbon nitride (g-C3N4) as a new drug delivery system for carboplatin to treat cancer. J. Mol. Liq. **331**, 115607 (2021). https://doi.org/10.1016/j.molliq.2021.115607
17. B. Makiabadi, M. Zakarianezhad, R. Behjatmanesh-Ardakani, S. H. Mousavi, Investigating the performance of BN nanotubes as drug delivery systems for Azacitidine and Decitabine anti-cancer drugs: A theoretical study. Comput. Theor. Chem., **1231**(2023), p. 114429, (2024). https://doi.org/10.1016/j.comptc.2023.114429
18. C. Morell, A. Grand, A. Toro-Labbé, New dual descriptor for chemical reactivity. J. Phys. Chem. A **109**(1), 205–212 (2005). https://doi.org/10.1021/jp046577a

19. M.B. Singh, P. Jain, F. Mohammad, P. Singh, I. Bahadur, O.P. Abedigamba, Significant increase in the dipole moment of graphene on functionalization: DFT calculations and molecular dynamics simulations. ACS Omega **9**(14), 16458–16468 (2024). https://doi.org/10.1021/acsomega.4c00173
20. J. Y. Chen, H. Jiang, S. J. Chen, C. Cullen, C. M. Sabbir Ahmed, Y. H. Lin, Characterization of electrophilicity and oxidative potential of atmospheric carbonyls. Environ. Sci. Process. Impacts, **21**(5), pp. 856–866, (2019). https://doi.org/10.1039/c9em00033j
21. R.G. Parr, R.G. Pearson, Absolute hardness: companion parameter to absolute electronegativity. J. Am. Chem. Soc. **105**(26), 7512–7516 (1983). https://doi.org/10.1021/ja00364a005
22. M. El Faydy et al., In situ synthesis, electrochemical, surface morphological, UV–visible, DFT and Monte Carlo simulations of novel 5-substituted-8-hydroxyquinoline for corrosion protection of carbon steel in a hydrochloric acid solution. J. Mol. Liq. **280**, 341–359 (2019). https://doi.org/10.1016/j.molliq.2019.01.105
23. J.L. Gázquez, Perspectives on the density functional theory of chemical reactivity. J. Mex. Chem. Soc. **52**(1), 3–10 (2008)
24. A. Hussain *et al.*, J. Pre-Proof. (2019)
25. O.C. Adekoya, G.J. Adekoya, E.R. Sadiku, Y. Hamam, S.S. Ray, Application of DFT calculations in designing polymer-based drug delivery systems: an overview. Pharmaceutics, **14**(9), (2022). https://doi.org/10.3390/pharmaceutics14091972
26. N. Andrýsková, J. Motyčka, M. Babincová, P. Babinec, M. Šimaljaková, Computational design of a novel dithranol-salicylic acid antipsoriatic prodrug for esterase-activated topical drug delivery. Appl. Sci. **14**(3), 1094 (2024). https://doi.org/10.3390/app14031094
27. S. Shahab, M. Sheikhi, F. Azarakhshi, DFT Study of physisorption effect of the curcumin on. pp. 0–6, (2019). https://doi.org/10.14102/j.cnki.0254-5861.2011-2052
28. R. Wang et al., Band structure engineering enables to UV-Visible-NIR photocatalytic disinfection: Mechanism, pathways and DFT calculation. Chem. Eng. J. **421**(P1), 129596 (2021). https://doi.org/10.1016/j.cej.2021.129596
29. I.S. Burmistrov, Y. Gefen, D.S. Shapiro, A. Shnirman, Mesoscopic stoner instability in open quantum dots: suppression of coleman-weinberg mechanism by electron tunneling. Phys. Rev. Lett. **124**(19), 196801 (2020). https://doi.org/10.1103/PhysRevLett.124.196801
30. S. S. Gloria Kang GJ, Ewing-Nelson SR, Mackey L, Schlitt JT, Marathe A, Abbas KM, "乳鼠心肌提取 HHS Public Access. Physiol. Behav. **176**(1), pp. 139–148, (2018) https://doi.org/10.1021/jacs.8b04455.Experimental
31. M. Rizwan, A. Ali, Z. Usman, N.R. Khalid, H.B. Jin, C.B. Cao, Structural, electronic and optical properties of copper-doped SrTiO3 perovskite: A DFT study. Phys. B Condens. Matter **552**, 52–57 (2019). https://doi.org/10.1016/j.physb.2018.09.022
32. A.R. Benrekia, N. Benkhettou, A. Nassour, M. Driz, M. Sahnoun, S. Lebègue, Structural, electronic and optical properties of cubic SrTiO 3 and KTaO 3: Ab initio and GW calculations. Phys. B Condens. Matter **407**(13), 2632–2636 (2012). https://doi.org/10.1016/j.physb.2012.04.013
33. G.F.S., The volume changes of five gases under high pressures. J. Franklin Inst., **197**(1), p. 98, 1924. https://doi.org/10.1016/s0016-0032(24)90498-4
34. H.A. Rahnamaye Aliabad, V. Hesam, I. Ahmad, I. Khan, Electronic band structure of LaCoO3/Y/Mn compounds. Phys. B Condens. Matter. **410**(1), pp. 112–119. (2013). https://doi.org/10.1016/j.physb.2012.11.002
35. O.M. Løvvik, Prytz, Density-functional band-structure calculations for La-, Y-, and Sc-filled CoP3-based skutterudite structures. Phys. Rev. B - Condens. Matter Mater. Phys., **70**(19), pp. 1–6. (2004). https://doi.org/10.1103/PhysRevB.70.195119
36. J. Xue, X. Guo, X. Wang, Y. Xiao, Density functional theory studies on cytosine analogues for inducing double-proton transfer with guanine. Sci. Rep. **10**(1), 9671 (2020). https://doi.org/10.1038/s41598-020-66530-8
37. S. Ataei, E. Nemati-Kande, A. Bahrami, Quantum DFT studies on the drug delivery of favipiravir using pristine and functionalized chitosan nanoparticles. Sci. Rep. **13**(1), 1–15 (2023). https://doi.org/10.1038/s41598-023-49298-5

38. A.P. Kusuma et al., Energetics of carboxylic acid-pyridine heterosynthon revisited: A computational study of intermolecular hydrogen bond domination on phenylacetic acid-nicotinamide cocrystals. Open Chem. **20**(1), 949–957 (2022). https://doi.org/10.1515/chem-2022-0207
39. M. Medimagh et al., Investigations on the non-covalent interactions, drug-likeness, molecular docking and chemical properties of 1,1,4,7,7- pentamethyldiethylenetriammonium trinitrate by density-functional theory. J. King Saud Univ. - Sci. **35**(4), 102645 (2023). https://doi.org/10.1016/j.jksus.2023.102645
40. H.O. Edet et al., Heteroatoms (B, N, S) doped quantum dots as potential drug delivery system for isoniazid: insight from DFT, NCI, and QTAIM. Heliyon **9**(1), e12599 (2023). https://doi.org/10.1016/j.heliyon.2022.e12599
41. F. Cort, R.F.W. Bader, Complementarity of QTAIM and MO theory in the study of bonding in donor—acceptor complexes. **249**, pp 633–662, (2005). https://doi.org/10.1016/j.ccr.2004.08.022
42. T. Lu, F. Chen, Multiwfn: A multifunctional wavefunction analyzer. J. Comput. Chem. **33**(5), 580–592 (2012). https://doi.org/10.1002/jcc.22885
43. F.H. Al-Ostoot et al., Structural and theoretical exploration of a multi-methoxy chalcone: Synthesis, quantum theory, electrostatics, molecular packing, DFT analysis, and in-silico anti-cancer evaluation. Heliyon **10**(13), e33814 (2024). https://doi.org/10.1016/j.heliyon.2024.e33814
44. P. Kumar, A. Arya, Ramachandran plot - A simplified approach. Pathfind. Res. Train. Found. India, p. 6, (2018). Available: https://www.researchgate.net/publication/330158666_Ramachandran_plot-_A_simplified_approach
45. S.A. Hollingsworth, P.A. Karplus, A fresh look at the Ramachandran plot and the occurrence of standard structures in proteins. Biomol. Concepts **1**(3–4), 271–283 (2010). https://doi.org/10.1515/bmc.2010.022
46. D.P.L. Renj et al., Density functional theory calculations, vibrational spectral analysis and topological analysis of 1-acethyl-2(4-isopropoxy-3-methoxyphenyl) cyclopropane with docking studies. Chem. Phys. Impact, **8** (2024). https://doi.org/10.1016/j.chphi.2024.100524
47. M.A. Khan, J. Iqbal, M. Ilyas, A.R. Ayub, Y. Zhu, H. Li, Controlled supramolecular interaction to enhance the bioavailability of hesperetin to targeted cancer cells through graphyne: A comprehensive in silico study. RSC Adv. **12**(10), 6336–6346 (2022). https://doi.org/10.1039/d1ra09112c
48. S. Riaz, K. Jaffar, M. Perveen, A. Riaz, S. Nazir, J. Iqbal, Computational study of therapeutic potential of phosphorene as a nano-carrier for drug delivery of nebivolol for the prohibition of cardiovascular diseases: a DFT study. J. Mol. Model., **27**(10), (2021). https://doi.org/10.1007/s00894-021-04907-w.
49. C. Berthomieu, R. Hienerwadel, Fourier transform infrared (FTIR) spectroscopy. Photosynth. Res. **101**(2–3), 157–170 (2009). https://doi.org/10.1007/s11120-009-9439-x
50. Y. Hu, F. Xue, Z. Zhang, The influence of thermal modification on lignite structure by FT-IR. Therm. Sci. Eng. **5**(1), 94 (2022). https://doi.org/10.24294/tse.v5i1.1532
51. T. Yu, Y. Li, X. Gu, Q. Li, Development of a hyaluronic acid-based nanocarrier incorporating doxorubicin and Cisplatin as a pH-sensitive and CD44-targeted anti-breast cancer drug delivery system. Front. Pharmacol. **11**(August), 1–11 (2020). https://doi.org/10.3389/fphar.2020.532457

Chapter 6
DFT in Solid-State Physics

Prakash Pandey and Sudhir K. Pandey

6.1 Introduction

Density Functional Theory (DFT) is a widely used and highly effective method that has revolutionized condensed matter physics and materials science by enabling the study of material properties at the atomistic and electronic levels. By describing the interaction of electrons within the framework of quantum mechanics, no other approach except DFT describes a wide range of physical properties in materials at a minimum computational cost. The core principle is that the ground-state properties of a stationary many-body system may be expressed only via the ground-state density ($\rho(\mathbf{r})$). In DFT, $\rho(\mathbf{r})$ is the basic quantity where it is a function of only three spatial coordinates. Therefore, it is computationally viable for the many-particle wave function, which contains $3N$ coordinates for the N-particle system. The basic formulation of DFT is based upon the two theorems of Hohenberg and Kohn (HK) [1] and Kohn-Sham equations (KS) [2]. KS equation is the one-particle Schrödinger equation of a non-interacting KS system generating the same $\rho(\mathbf{r})$ as in the case of interacting inhomogeneous systems [2]. These are a set of self-consistent equations based on the formalism as provided by HK [1]. These equations mainly consist of external potential, Hartree potential, and exchange-correlation (XC) potential. To date, the XC potential has no exact form, which is responsible for explaining the quantum mechanical features of fermions. The quest for exact XC functionals has been a significant challenge in DFT for a long time. Two important families of

P. Pandey (✉)
School of Physical Sciences, Indian Institute of Technology Mandi, Kamand, India
e-mail: prakashpandey6215@gmail.com

S. K. Pandey
School of Mechanical and Materials Engineering, Indian Institute of Technology Mandi, Kamand, India
e-mail: sudhir@iitmandi.ac.in

R. Thotagamuge (ed.), *Density Functional Theory*,
https://doi.org/10.1007/978-981-95-9745-1_6

XC functionals, namely the local-density approximation (LDA) and the generalized-gradient approximation (GGA), are generally used to study the physical properties of materials. However, they do not properly capture the effect of electronic correlation. Also, it is known that DFT fails to provide accurate information about the band gap of strongly correlated electron systems because it improperly considers the localization effect of *d*/*f* orbitals. The most common way to get rid of this problem is to go beyond the DFT method, such as DFT+U, G_0W_0, GW_0, and DFT+DMFT. These methods provide fairly accurate results and reproduce the experiments quite well.

This chapter aims to explore the applications of DFT in materials science, focusing on crystalline materials, band gap calculations, lattice dynamics, and to demonstrate the extent to which DFT is useful in explaining the band theory of electrons in crystals. We will start by introducing the estimation of physical properties of various crystalline materials, followed by a discussion on determining crystal structures and their stability using different theoretical methods. Next, we will focus on band gap calculations of crystalline materials and compare the band gaps obtained using various XC functionals. We shall then discuss the general theory of lattice vibration and its application. After these preliminary steps, we will also discuss how to calculate the phonons and lattice dynamics properties through various DFT packages.

6.2 Crystalline Materials and the Application of DFT

6.2.1 Crystal Structure and the Concept of a Brillouin Zone

A crystal structure gives a complete description of the arrangement of atoms, molecules, or ions in a periodic lattice, which makes up the basic building blocks of a material [3]. The Bravais lattice just describes the geometry of the underlying periodic structure, independent of the actual units, which might be single atoms, groups of atoms, molecules, ions, etc. Based on the symmetry operations of the unit cell, the 14 Bravais lattices are divided into the triclinic, monoclinic, orthorhombic, tetragonal, rhombohedral, hexagonal, and cubic lattice systems [3]. The first Brillouin zone, defined as the Wigner-Seitz primitive cell of the reciprocal lattice that contains all the wave vectors (**k**). Wigner-Seitz cell and first Brillouin zone refer to identical geometrical constructions. The smallest region completely surrounded by planes that are the perpendicular bisectors of the reciprocal lattice vectors taken from the origin is known as the first Brillouin zone (BZ). There are other BZs besides the first one, such as the second, third, and so on, but their usage is less common. In the calculation of electronic spectrum, one may often examine the eigenvalues and eigenstates at high symmetric k-points, to investigate both the trivial and non-trivial features present in the systems, including the trivial and non-trivial (bulk) energy gap and shape of the bands near the Fermi energy. In DFT calculations, the first BZ plays a vital role by accessing the direct and indirect band gap, impacting the optical, electronic, magnetic, and topological properties of solids.

Limited computational resources limit the search for novel materials and their exotic properties. DFT, a popular and potential method for determining the physical properties of materials, requires integration over the energy spectrum of a material. It uses the concept of the BZ for studying the physical properties of the material. Also, the irreducible part of BZ requires fewer computational resources than the full BZ. By taking full advantage of the irreducible part of BZ, DFT can calculate various material properties that are governed by their electronic structure, including thermoelectricity, thermophysical properties, optical properties, non-trivial topological properties, and superconductivity.

The Periodic Potential and Bloch's Theorem: In a perfect crystal, the ions are arranged in a regular periodic fashion. Considering the problem of an electron in a potential $U(\mathbf{r})$ with the periodicity of Bravais lattice is given by[3]:

$$U(\mathbf{r}+\mathbf{R}) = U(\mathbf{r}) \tag{6.1}$$

Here $\mathbf{R}$ is Bravais lattice vectors. This periodicity plays a crucial role in predicting the physical properties of the crystalline materials. Since the periodicity of the potential U is about 1 Å, which corresponds to the size of an electron's de Broglie wavelength in the Sommerfeld free electron model [3]. This implies that the influence of periodicity on electron motion can only be explained using quantum mechanics. It is important to note that perfect periodicity is an idealization. Real solids are never absolutely pure, and the presence of impurity atoms causes the solid to differ locally from the rest of the crystal. Moreover, even in an absolutely pure crystal, there is always a slight, temperature-dependent probability of missing or misplaced ions, which disrupts perfect translational symmetry [3]. The above-mentioned imperfections are highly important. They are, for example, ultimately responsible for the fact that the electrical (thermal) resistivity of metals is not zero (infinite) [3]. Furthermore, it is important to emphasize that all crystalline solids encounter the problem of electrons in a periodic potential.

The problem of electrons in a solid is a many-electron problem, which is one of the most intractable problems of quantum physics. This is because the full many-body Hamiltonian of the solids encompasses all possible interactions, such as electron-electron, electron-ion, and ion-ion interactions. In the independent electron approximation, these interactions are represented by an effective one-electron potential $U(\mathbf{r})$. Regardless of the form of U, if the crystal is perfectly periodic, it must satisfy relation (6.1). Independent electrons, each obeying one-electron Schrödinger equation with a periodic potential, are known as Bloch electrons. The stationary states of Bloch electrons possess a crucial property due to the periodicity of the potential U.

Bloch's Theorem, a cornerstone of condensed matter physics and materials science, facilitates the calculation of electronic wave functions and energies by solving Schrödinger's equation. The theorem, named after the Swiss physicist Felix Bloch, is stated as follows [3]:

Theorem 1 *The eigenstates ψ of the one-electron Hamiltonian $H = -\hbar^2\nabla^2/2m + U(\mathbf{r})$, where $U(\mathbf{r}+\mathbf{R}) = U(\mathbf{r})$ for all $\mathbf{R}$ in a Bravais lattice, can be chosen to have the form of a plane wave times a function with the periodicity of the Bravais lattice:*

$$\psi_{nk}(\mathbf{r}) = e^{ik\cdot r} u_{nk}(\mathbf{r}) \tag{6.2}$$

where

$$u_{nk}(\mathbf{r}) = u_{nk}(\mathbf{r}+\mathbf{R}) \tag{6.3}$$

for all **R** in a Bravais lattice. From Eqs. 6.2 and 6.3, one can write

$$\psi_{nk}(\mathbf{r}+\mathbf{R}) = e^{ik\cdot r} \psi_{nk}(\mathbf{r}) \tag{6.4}$$

Also, one can write Bloch's theorem in this alternative form: the eigenstates of H can be chosen so that associated with each is a wave vector **k** such that

$$\psi(\mathbf{r}+\mathbf{R}) = e^{ik\cdot r} \psi(\mathbf{r}) \tag{6.5}$$

This theorem suggests that, instead of considering an infinite crystal, one can focus on the unit cell, thereby reducing the complexity of computations. Also, it is the fundamental in all first-principles approaches to calculate the various properties of crystalline material. In DFT calculations, researchers calculate band gaps, electron mobility, mechanical stability, and many other properties by employing this theorem. Moreover, this can also provide the room for the analysis of insight phenomena that are due to the influence of electrons in a periodic lattice [3].

6.2.2 Application: Determining Crystal Structures and Predicting Stability

Understanding the behavior and properties of materials is highly important for many technological applications. Predicting the crystal structure is a crucial task in materials sciences, as it links many physical properties and their fascinating phenomena. However, the search for a stable crystal structure remains computationally challenging, as it requires identifying the atomic configuration with the lowest total energy among numerous possibilities. The prediction of the crystal structure at zero Kelvin using quantum mechanical methods has become feasible due to the substantial advancements in computational resources. Nowadays, there are many ways to predict the crystal structure correctly. One method is to extract known structures from databases of structures previously found in similar materials. However, the success rate of this method is very limited and fails to generate a new crystal structure. There are many more advanced methods that have been proposed, such as simulated annealing [4], minima hopping [5], basin hopping [6], metadynamics [7], the genetic algorithm [8], and the random sampling method [9]. Despite the successes, the predictions are found to be incorrect in many cases. Therefore, for correct crystal structure prediction, the researchers have used evolutionary algorithms (e.g., USPEX code) [10] and particle swarm optimization (e.g., CALYPSO code) [11].

USPEX-Evolutionary Crystal Structure Prediction: The fascinating feature of USPEX is that it can find the most stable crystal structure of systems with up to several dozen atoms/cell [10, 12]. The algorithms used in this code also found highly competitive metastable structures through local optimization and the process of exploiting promising regions. Moreover, this code also identifies the crystal structure at different environmental conditions such as pressure and temperature. Instead of identifying stable phases, this code can thus be used for materials design, both in finding promising structures to synthesize and providing the information on the best suited conditions for synthesizing the material. To illustrate the capabilities of this code, we have included examples in a Table 6.1 where the predicted crystal structures are compared with reported structures obtained either theoretically through first-principles methods or experimentally synthesized.

Examples of Predicting Crystal Structures Through USPEX: USPEX uses evolutionary algorithms to predict the crystal structure under different conditions, such as temperature and pressure [10]. For example, under high pressure, $CaCO_3$ exhibits a phase transition, which is predicted by this code and verified by the reported crystal structure. The structure found by USPEX closely reproduces the experimental diffraction pattern. In agreement with the experiment, calculations show that post-aragonite structure ($P2_12_12$) is more stable than aragonite structure (Pmcn) at 50 GPa [10]. Also, the result obtained for post-aragonite structure is mentioned in Table 6.1. Above 137 GPa, the code predicts the stability of another new structure, $C222_1$ [10], which has been verified experimentally. Another example is $MgSiO_3$, which is a major Earth-forming compound, comprising approximately 40% of the Earth's volume or about 80% of its lower mantle. This code predicted the Pbnm phase at 120 GPa, while at zero pressure, it belongs to the $Fm\bar{m}3m$ crystal structure. The results were also compared with the reported findings.

Table 6.1 The predicted crystal structure at different pressure magnitudes, obtained using the USPEX code, along with other reported results, is presented in the table

Systems	Pressure (GPa)	Predicted structures	Reported results
$CaCO_3$	50	$P2_12_12$ [10]	$P2_12_12$ [13]
	86	$P2_12_12$ [10]	$P2_12_12$ [13]
	150	$C222_1$ [10]	$C222_1$ [14]
$MgSiO_3$	120	Pbnm [10]	Pbnm [15, 16]
	0	$Fm\bar{m}3m$ [12]	$Fm\bar{m}3m$ [17]
Al	162	P6(3)/mmc [12]	P6(3)/mmc [17]

CALYPSO: Particle Swarm Optimization Crystal Structure Prediction: The CALYPSO code employs particle swarm optimization (PSO), a global minimization method, to predict crystal structures. This code primarily involves four steps: generating random structures constrained by symmetry, performing local structural optimization, postprocessing to identify unique local minima using geometric structural parameters, and generating new structures via PSO for subsequent iterations. The CALYPSO code can predict various structures for elemental, binary, and ternary compounds across diverse chemical bonding environments (e.g., metallic, ionic, and covalent bonding). Additionally, this code can discover unknown structures under different environmental conditions, such as varying pressure. To demonstrate the capabilities of this code, we have included examples in the Table 6.2. This table compares the predicted crystal structures with those reported theoretically (using first-principles methods) or synthesized experimentally.

Examples of Predicting Crystal Structures Through CALYPSO: Lithium is a simple alkali metal at ambient pressure, but under compression, it exhibits phase transitions from BCC→FCC→hR1→cI16. Beyond these phases, experimental exploration has been limited. However, the CALYPSO code predicts stable structures at 0, 10, 40, 70, 80, 100, 120, 200, and 300 GPa [11], revealing new phases, as mentioned in Table 6.2. Another element, magnesium, was also tested, and the simulations successfully reproduced all the experimental structures. This highlights the effectiveness of this approach in predicting crystal structures. The CALYPSO code not only predicts structures at the elemental level but also determines the crystal structures of compounds under varying pressures. For example, this code predicts the structures of Bi_2Te_3 at different pressures. At 12 GPa, Bi_2Te_3 belongs to the β-phase, at 14 GPa to the γ-phase, and at 20 GPa to the C2/m structure. These phase transitions are further validated by other reported results, as shown in the table.

Table 6.2 The predicted crystal structure at different pressure magnitudes, obtained using the CALYPSO code, along with other reported results, is presented in the table

Systems	Pressure (GPa)	Predicted structures	Reported results
Li	80	Aba_2 [11]	Aba_2 [18]
	200	Cmca [11]	Cmca [18]
Mg	500	fcc [11]	fcc [19]
	800	sh [11]	sh [19]
Bi_2Te_3	12	β-Bi_2Te_3 [11]	β-Bi_2Te_3 [20]
	14	γ-Bi_2Te_3 [11]	γ-Bi_2Te_3 [20]
	20	C2/m [11]	C2/m [20]
BC_3	0	Pmma [11]	Pmma [21]
BC_7	0	P-4m2 [11]	P-4m2 [22]

6.2.3 Calculation of Band Gaps

The band gap of semiconductor and insulator is the most important parameter for a wide range of applications, including solar cells, high-temperature and power switching, power electronics, displays, sensors, and lasers. It is a fundamental property of semiconductors and insulators, essential for understanding their optical and electrical properties. The fundamental band gap of a semiconductor and insulator is defined as the difference between its ionization potential and electron affinity. However, accurately predicting band gaps using DFT, which is crucial for discovering novel materials, remains a challenging task. DFT calculations based on LDA (GGA) are often used to predict stable crystal structures, but they typically underestimate (overestimate) the experimental volume [23, 24]. In addition to this, when LDA or GGA functionals are used, the band gaps predicted by DFT are significantly underestimated compared to experimental values. To improve band gap, hybrid functionals are used, offering more accuracy than LDA and GGA. However, predicting the band gap of insulators requires going beyond LDA or GGA approximations in DFT, significantly increasing the computational cost. One of the advantages of DFT is that the many-electron problem is addressed using the independent-electron approximation, which provides reasonably good agreement with experimental data for various physical properties of many materials. However, DFT is known to fail in accurately describing the electronic and magnetic properties of materials containing partially filled d and/or f-block elements, commonly referred to as strongly correlated electron systems [25]. Additionally, obtaining information about excited-state spectra for any material is challenging using DFT alone. Several methods have been proposed to overcome these limitations. One such method is DFT+U, which provides significantly improved predictions of the insulating ground state for many materials, aligning well with experimental observations, whereas standard DFT often predicts a metallic ground state for the same materials. However, some spectroscopic observations for f-electron systems are not adequately explained by the DFT+U method. Also, in DFT+U, U is taken as a parameter. So, it is a natural question to ask what the suitable value of U is for any material to explain the band gap problem. Hence, this method loses its predictive power in the case of experimentally unexplored materials. Therefore, a better electronic theory is needed to address band-gap problems. Nowadays, the GW approximation, based on many-body perturbation theory, is recognized as a parameter-free electronic structure method. In various studies, it has been observed that the GW method can yield very accurate band gaps [26–30], but it requires a very high computational cost, particularly if done self-consistently [27, 29]. Therefore, one requires a computationally cheaper theory than GW. From the perspective of the band gap, it has been reported that hybrid functionals largely overcome the problem of the band gap [31, 32]. However, hybrid functionals are more computationally expensive than traditional (semi)local DFT functionals [33]. It has also been reported that hybrid functionals overestimate the equilibrium volume and underestimate the atomization energies [34]. Therefore, a more advanced and computationally cheaper theory is needed to explain the physical

Table 6.3 The Band-Gap Energies (eV) of Metal Oxides, Chalcogenides, and Nitrides Using Various Density Functionals, in Comparison with Experimental Data and the GW Method are presented. TB-mBJ results using Wien2k are given. The value of U is 5 eV. (A) and (R) indicate the Anatase phase and Rutile phase of the compounds, respectively. Except for the GW and experimental band gap values, the band gaps of all the systems are taken from Ref. [38]

Systems	PBE	GGA+U	PBE0	B3LYP	HSE06	TB-mBJ	GW value	Exp value
TiO_2(A)	1.94	2.92	4.11	4.03	3.38	3.05	3.56 [39]	3.30 [40]
TiO_2(R)	1.64	2.48	3.87	3.46	3.12	2.67	3.34 [39]	3.03 [41]
ZnO	0.87	1.42	3.18	3.30	2.54	2.68	3.20 [42]	3.30 [43]
FeO	0.0	1.97	3.00	2.97	2.25	1.83	2.20 [44]	2.40 [45]
NiO	0.88	3.01	5.19	4.93	4.45	4.14	3.80 [46]	4.20 [47]
V_2O_5	1.74	2.73	3.98	3.57	3.23	2.87	1.85 [48]	2.30 [49]
WO_3	1.79	2.31	3.68	2.50	2.94	2.90	2.93 [50]	2.80 [51]
Bi_2O_3	2.45	2.44	4.11	3.90	3.50	3.36	–	2.85 [52]
MoS_2	0.90	0.93	2.11	1.93	1.46	1.11	1.23 [53]	1.23 [54]
WS_2	1.07	1.11	2.25	2.00	1.60	1.31	1.30 [53]	1.35 [54]
$MoSe_2$	0.91	0.87	2.03	–	1.36	1.02	1.11 [53]	1.09 [54]
WSe_2	0.99	0.99	2.06	–	1.44	1.20	1.19 [53]	1.20 [54]
GaN	2.07	2.31	3.87	3.79	3.22	3.10	3.35 [55]	3.50 [56]

properties of materials. It is reported that among the affordable band gap correction schemes [35, 36], the TB-mBJ potential provides reasonably accurate band gap predictions with the same computational cost as LDA/GGA functionals [37]. Therefore, it is concluded that the TB-mBJ method competes in accuracy with the expensive hybrid and GW methods while being efficiently applicable to very large systems.

Let us now examine the issues with band-gap predictions through DFT and explore how various functionals and beyond-DFT theories address these problems. For this, we have selected examples of different compounds, as mentioned in Table 6.3. In this table, the band gaps of a wide variety of materials, including metal oxides, chalcogenides, and nitrides, along with the resulting band-gap sizes, are presented. It is well known that GGA generally underestimates the band gaps, with errors ranging from ~2% for dichalcogenides to ~77% for rock-salt type transition-metal oxides. For a clearer understanding of issues with band-gap predictions, we take the example of FeO, which is a Mott insulator [57]. This material is predicted to be metallic by the PBE XC functional due to its failure to properly capture the *d*-orbitals, as mentioned in Table 6.3. The metallic behavior in this material is well known for LDA and GGA functionals [58]. The situation improves with the use of Hubbard U=5 eV, which is sufficiently large, yielding a band gap of 1.97 eV. However, this still underestimates the experimental band gap (2.40 eV) [45]. It is important to mention that to achieve the experimental band gap, one can tune the U values. However, different U values are used for the same material to study various physical properties. Hence, this creates a problematic situation for the theoretical prediction of new materials. The results indicate that a more advanced theoretical approach is needed to achieve the experimental band gap. For this, the use of the hybrid PBE0 functional provides a slight improvement (3.0 eV), but this functional tends to overestimate band gaps. The errors decrease by up to 50% compared to pure GGA functionals. The advantage of this functional is that Mott insulators are correctly converged to an insulating state. Similarly, the other hybrid functional, B3LYP (HSE06), gives band-gap energies similar (different) to those of PBE0, exhibiting the same overestimating (underestimating) behavior. Moreover, the band gap obtained using the HSE06 functional (2.25 eV) is in close agreement with the experimental value. However, HSE06 functional is a slower method than traditional (semi)local DFT functionals [33]. In addition to this, hybrid functionals fail to accurately account for the properties of metals, vastly overestimating the equilibrium volume and underestimating the atomization energies [34]. To accurately describe all aspects, a more advanced functional is required. Li et al.. [38] reported a band gap of 1.83 eV using the TB-mBJ functional, which yields band gap energies close to those of GGA+U. In general, the TB-mBJ potential can yield highly accurate band gaps for most semiconductors and insulators [59]. For example, in NiO, which is a strongly correlated transition-metal compound, TB-mBJ predicts a band gap of 4.14 eV, which is very close to the experimental value of 4.20 eV [47]. Furthermore, in addition to the band structure, other properties such as the magnetic moment and electron density can be accurately described by the TB-mBJ potential [59]. Moreover, this method is computationally cheap while offering a quantitative predictive power comparable to, or sometimes even better than, more sophisticated and costly methods [60]. In addition, to obtain both the ground-state and quasiparticle energies correctly within a single formalism, the GW method can be used. From the table, it can be seen that using the GW method, the band gap for FeO obtained (2.20 eV) is very close to the experimental value. However, it is much more computationally demanding compared to the approach we discussed earlier, which provides essentially the same results. Therefore, it is concluded that the TB-mBJ method provides an excellent description of the band gaps of semiconductors and insulators at a lower computational cost.

6.3 Phonons and Lattice Dynamics in Crystalline Solids

It is convenient to visualize an atom in a crystalline solid as consisting of its ion core and valence electrons. The ion cores vibrate about their equilibrium positions, while the valence electrons move around these ion cores. In a crystalline solid, the motion of all ion cores and electrons can be simplified by leveraging the lattice's translational periodicity. This periodicity effectively reduces the problem to the dynamics of the atoms within the primitive unit cell, which contains the ion cores and their associated electrons. However, providing a detailed description of even this "small" system remains a challenging task. To simplify the problem further, two key approximations are introduced. Since ion cores are much heavier than electrons, their motion can be treated separately—a concept known as the *adiabatic approximation*. As a consequence of this approximation:

1. When analyzing the motion of electrons, it is assumed that the ion cores remain fixed at their equilibrium positions.

2. The motion of the ion cores is determined within a potential field generated by the average motion of the electrons.

It is important to note that the second assumption is one of the most challenging aspects of a first-principles study of lattice dynamics, which we will discuss in greater detail later.

Phonons play a crucial role in understanding the vibrational properties of elements and compounds. The phonon density of states provides the spectrum of phonon energies, which is essential for calculating the thermodynamic properties of materials. To analyze and compute the thermodynamic properties of solids, such as free energy, entropy, and heat capacity, harmonic phonons offer a valuable framework. Harmonic phonons are non-interacting elementary excitations characterized by their frequencies and possess infinite lifetimes [3]. Within the harmonic approximation, individual phonons are treated as independent quantum mechanical entities. A wide range of important quantum phenomena arise from interactions beyond the harmonic approximation, such as electron-phonon interactions [61] and phonon-phonon interactions [62]. These interactions are associated with finite phonon lifetimes and phonon frequency shifts. Fundamental properties of crystals, such as electrical resistivity and superconductivity, are governed by electron-phonon interactions. Phonon-phonon interactions, on the other hand, originate from the intrinsic anharmonicity of the crystal lattice and contributes to a finite lattice thermal conductivity. Additionally, various novel phenomena, including phonon drag [3] and hydrodynamic phonon transport [63], are also influenced by phonon-phonon interactions.

6.3.1 Phonon Calculations

Phonon calculations are commonly performed using two primary methods: the finite-displacement supercell method (also known as the frozen phonon approximation) and density-functional perturbation theory (DFPT) [64, 65]. In the finite-displacement supercell method, a supercell of the crystal is first generated. Using first-principles methods, the atomic forces are calculated by displacing an atom from its symmetrically allowed position. These forces are then used to determine the force constants, which serve as the foundation for phonon calculations.

In contrast, the DFPT method, based on ground-state electronic structure calculations, takes into account perturbations in the electronic density and potential energy. These perturbations arise from small atomic displacements from their equilibrium positions in different directions and are incorporated into the Hamiltonian of the Kohn-Sham equations. The changes in the electronic structure due to these perturbations are used to calculate the force constants, which are subsequently employed to construct the dynamical matrix and carry out phonon calculations.

Phonon calculations involve determining the eigenvalues (phonon frequencies) and eigenvectors (phonon modes) of a system. In these calculations, the solid is modeled under the Born-von Kármán periodic boundary condition [3], where the mean position of each atom follows periodic constraints. In this model, an atom is displaced from its mean position, and the resulting forces on all atoms are calculated. These forces are then used to compute the force constants, which are essential for constructing the dynamical matrix for any arbitrary phonon wave vector **q**. By diagonalizing the dynamical matrix, the eigenvalues (phonon frequencies) and eigenvectors (phonon modes) are obtained. There are two primary methods for performing phonon calculations, which are discussed in subsequent sections. Several first-principles packages, such as WIEN2K [66], Elk [67], ABINIT [68], Quantum ESPRESSO [69], VASP [70], CASTEP [71], SIESTA [72], and LAMMPS [73], are interfaced with Phonopy [74], which is used for phonon calculations through the supercell approach. Meanwhile, Elk, ABINIT, Quantum ESPRESSO, and VASP use both DFPT and the supercell approach to calculate the phonon dispersion.

Supercell Approach: In the supercell approach, the net force on each atom within the unit cell is calculated based on its interactions with surrounding atoms. A larger supercell size allows more atoms to contribute to these interactions, resulting in more accurate force calculations as it better approximates an infinite crystal lattice. Studies have shown that choosing an appropriate supercell size provides a more accurate representation of phononic properties. Additionally, the dependence of phononic properties on supercell size can be understood through various technical aspects. For a fixed **k**-mesh size, the Brillouin zone sampling is determined by the size of the supercell. A smaller supercell leads to coarser Brillouin zone sampling, which may fail to capture critical features or introduce anomalies in the phonon dispersion. In contrast, a larger supercell offers finer Brillouin zone sampling, yielding a more accurate depiction of phonon properties. Furthermore, smaller supercells may produce inaccurate results due to zone-folding and finite-size effects. Now, let us examine a case study to gain a clearer understanding of the supercell approach.

BCC Vanadium and Niobium Metal: From Fig. 6.1, it is seen that both Vanadium (V) and Niobium (Nb) are mechanically stable for supercell size $2\times2\times2$. However, with increasing the supercell size to $3\times3\times3$, negative frequencies are foundÃ‚ in the small region of Γ-H and Γ-N directions. On further increase in supercell size to $4\times4\times4$, imaginary frequencies completely disappear in both elements. Furthermore, in a $4\times4\times4$ ($3\times3\times3$) supercell of V a small kink (very small kink) is observed in the vicinity of the Γ-point along Γ-H high symmetric directions. However, in the case of Nb, the size of kink is bigger than the V corresponding to $4\times4\times4$ supercell. Also, it is bigger in comparison to the $3\times3\times3$ supercell. In addition to this dip-like feature in the longitudinal acoustic phonon mode along the Γ-H high symmetric path is observed for V and Nb in the case of supercell size $4\times4\times4$. Also, in supercell size $3\times3\times3$, a small dip is also seen in the longitudinal acoustic phonon mode. But a small instability occurs in this supercell size along Γ-H and Γ-N high symmetric directions, which is spurious. However, in the supercell size $2\times2\times2$, dip-like feature

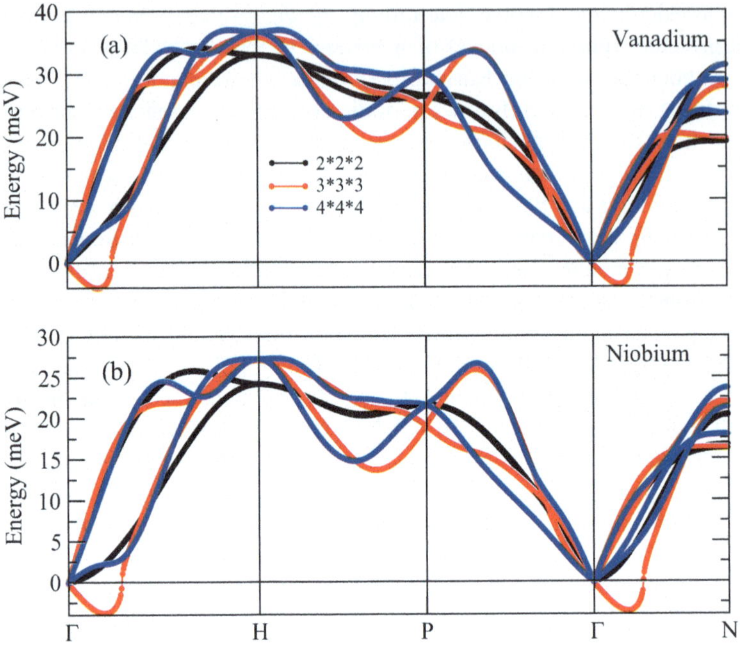

Fig. 6.1 (Color online) Phonon dispersion curves of BCC **a** Vanadium and **b** Niobium metals [75]. The black, red and blue curve represent phonon dispersion calculations at supercell 2×2×2, 3×3×3, and 4×4×4. These calculations were performed using the WIEN2K-Phonopy [74] package

is absent in both elements. The origin of the dip-like feature in the different sizes of supercells is due to long-range interactions, which play a crucial role in the phonon dispersion curve [3]. This dip-like feature in the phonon dispersion is termed as soft mode. From the obtained results, it is found that the soft mode mainly depends on the size of the supercell. We therefore conclude that the supercell size 4×4×4 is more accurate for the calculation of phononic properties of V and Nb.

From the above examples, we have highlighted the importance of the supercell size in calculating the phononic properties of materials. Additionally, to address the occurrence of imaginary frequencies during first-principles calculations, here are some methods that should always be kept in mind.

(1.) Before starting phonon calculations, one should optimize the atomic positions if they are variable. Additionally, it is important to optimize the lattice parameters corresponding to the functional being used.

(2.) After this, one should ensure the phonon calculations are converged with respect to the supercell size, as discussed earlier. Additionally, the obtained dispersion curves should also be converged with respect to the **k**-mesh size.

These are the key points that one should always keep in mind when performing phonon calculations.

DFPT Approach: DFPT is a very popular and efficient method to evaluate phonon frequencies and electron-phonon coupling matrix elements within DFT [76]. To calculate the phononic properties of a solid using this approach, one should optimize the crystal structure to the lowest energy state by minimizing all the forces on the atoms. Then one can calculate the total energy of the crystal by distorting the first atom from its equilibrium position through the Kohn-Sham DFT. Using the examples of V and Nb, let us now understand phonon calculations through the DFPT approach.

From Fig. 6.2, it is seen that calculations of phonon dispersion are performed at $2\times2\times2$, $4\times4\times4$, and $6\times6\times6$ q-mesh sizes through the DFPT approach. Figure 6.2 reveals that using a $2\times2\times2$ q-mesh results in phonon dispersion of both elements

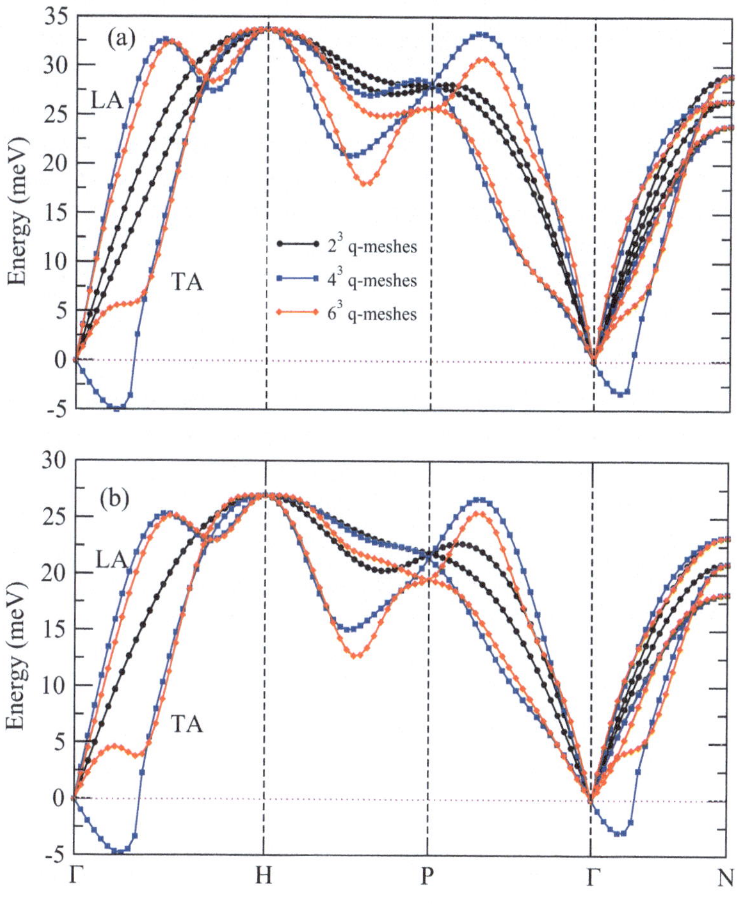

Fig. 6.2 (Color online) Phonon dispersion curves of BCC **a** Vanadium and **b** Niobium metals [75]. The black, red and blue curve represent phonon dispersion calculations at **q**-mesh $2\times2\times2$, $4\times4\times4$, and $6\times6\times6$. These calculations were performed using the ABINIT [68] package through the DFPT approach

with no imaginary frequencies. But the dip-like features in phonon dispersion of both elements are not found along Γ-H high-symmetry directions. However, increasing the q-mesh size to 4×4×4, imaginary frequencies are obtained along the small region of Γ-H and Γ-N high-symmetry directions, which is spurious. On further increase in q-mesh size to 6×6×6, imaginary frequencies completely disappear in both elements. Moreover, in this q-mesh size, the "dip"-like features are clearly visible, and the observed dip is the same as that found in the 4×4×4 supercell. Therefore, one should always ensure that the phonon dispersion is converged with respect to the q-mesh size. Additionally, the obtained dispersion curves should also be checked for convergence with respect to the **k**-mesh size.

6.3.2 Lattice Thermal Conductivity

Phonons play a significant role in studying many physical properties of solids, such as thermal conductivity [75], thermal expansion, and superconductivity. These properties have various technological and scientific applications, including thermoelectricity, heat management, and the investigations of the deep Earth. Understanding phononic transport properties has always been a challenging task due to the involvement of many-body interactions, such as electron-phonon, phonon-phonon, and phonon-defect interactions. In materials, these interactions have a considerable effect on transport properties, making the study even more challenging. However, advancements in computational resources have enabled us to compute phononic transport properties related to the anharmonicity of materials at finite temperatures. Anharmonic phonon calculations provide the imaginary part of the phonon self-energy, which is used to determine the phonon lifetime resulting from phonon-phonon interactions. This lifetime is further utilized to calculate lattice thermal conductivity, an important parameter in the application of thermoelectric materials.

Various first-principles packages, such as PHONO3PY [77, 78], ShengBTE [79], Phoebe [80], ALAMODE [81], and GULP [82], are used to calculate lattice ther-

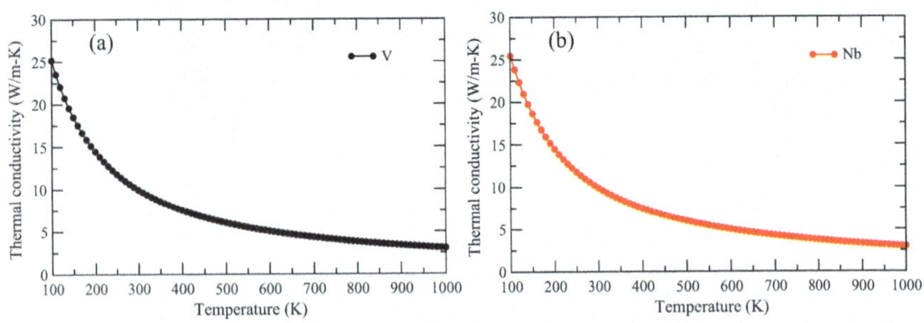

Fig. 6.3 (Color online) Lattice thermal conductivity **a** Vanadium and **b** Niobium metals as a function of temperature [75]. These calculations were performed using the PHONO3PY [77] package

mal conductivity. Typically, in these codes, the lattice thermal conductivity (κ_{ph}) is calculated using the single-mode relaxation time approximation of the linearized phonon Boltzmann equation. As mentioned above, the imaginary part of the phonon self-energy Γ_λ is a key factor in calculating κ_{ph}, and it is defined as

$$\Gamma_\lambda(\omega) = \frac{18\pi}{\hbar^2} \sum_{\lambda'\lambda''} |\Phi_{\lambda\lambda'\lambda''}|^2 \{(n_{\lambda'} + n_{\lambda''} + 1) \times \delta(\omega - \omega_{\lambda'} - \omega_{\lambda''}) + (n_{\lambda'} - n_{\lambda''}) \times [\delta(\omega + \omega_{\lambda'} - \omega_{\lambda''}) - \delta(\omega - \omega_{\lambda'} + \omega_{\lambda''})]\} \quad (6.6)$$

Here, $\Phi_{\lambda\lambda'\lambda''}$ denotes strength of interaction between three phonon λ, λ' and λ'' in the scattering. n_λ represents the Bose–Einstein distribution function at equilibrium and ω_λ is frequency of phonon mode. The phonon lifetime of mode λ is obtained as $\tau_\lambda = 1/2\Gamma_\lambda(\omega)$. Here, $2\Gamma_\lambda(\omega)$ is the phonon linewidth of mode λ. Under the single-mode relaxation time approximation, the κ_{ph} is obtained using linearized phonon Boltzmann equation as

$$\kappa_{\text{ph}} = \frac{1}{NV_0} \sum_\lambda C_\lambda \mathbf{v}_\lambda \otimes \mathbf{v}_\lambda \tau_\lambda^{\text{SMRT}} \quad (6.7)$$

where, N, V_0, and $\mathbf{v}_\lambda$ denote the number of unit cells, volume of a unit cell, and group velocity, respectively. C_λ and $\tau_\lambda^{\text{SMRT}}$ represent heat capacity and single-mode relaxation time of phonon mode λ. Let us now consider the case studies of V and Nb to understand lattice thermal conductivity.

In Fig. 6.3, the lattice thermal conductivity of V and Nb are shown within the temperature range 100–1000 K. As shown in figures, κ_{ph} decreases with increasing temperature. At 300 K, the calculated value of κ_{ph} for V (Nb) is 9.93 (9.95) W/m-K, while it reaches 3.05 (3.04) W/m-K at 1000 K [75]. As it is known, the thermal conductivity of materials, κ, is the sum of the electronic componentÃ,Â (κ_{el}) and phononic component (κ_{ph}), i.e., $\kappa = \kappa_{\text{el}} + \kappa_{\text{ph}}$. According to the Wiedemann-Franz law, κ_{el} of a material is directly related to electrical conductivity [3]. Similar to phonon dispersion curves, one should also converge lattice thermal conductivity with respect to the k-mesh size, as phonon transport properties are highly dependent on the k-mesh size.

6.4 Summary and Future Directions in DFT for Material Science Applications

Density Functional Theory (DFT) has emerged as an effective and powerful approach in condensed matter physics and materials science for calculating a wide range of physical properties of solids. In recent years, the field of materials science has witnessed enormous progress, with the creation of extensive materials datasets made possible largely due to the success of DFT and rapid advancements in computational capabilities. In the early days, challenges included calculating properties such as the band gap at finite temperatures, as calculations were initially limited to $T = 0$ K. Despite significant progress, many challenges remain for future research. One major hurdle is identifying an appropriate exchange-correlation functional capable of accurately describing all properties without requiring higher-level corrections. Such a functional should balance computational efficiency with accuracy and perform well across various applications.

Finally, materials science is transitioning into a new era of data-driven research. While relative success has been achieved, the construction of a broader and more comprehensive framework remains an ongoing process. The possibilities and limitations of this approach are only beginning to be understood by the scientific community. Furthermore, the ever-growing volume of scientific data continues to invite theoretical, computational, and experimental scientists to explore and leverage its potential.

References

1. P. Hohenberg, W. Kohn, Inhomogeneous electron gas. Phys. Rev. **136**, B864–B871 (1964)
2. W. Kohn, L.J. Sham, Self-consistent equations including exchange and correlation effects. Phys. Rev. **140**, A1133–A1138 (1965)
3. N.W. Ashcroft, N.D. Mermin, *Solid State Physics*. Cengage Learning (2022)
4. J. Pannetier, J. Bassas-Alsina, J. Rodriguez-Carvajal, V. Caignaert, Prediction of crystal structures from crystal chemistry rules by simulated annealing. Nature **346**(6282), 343–345 (1990)
5. S. Goedecker, Minima hopping: an efficient search method for the global minimum of the potential energy surface of complex molecular systems. J. Chem. Phys. **120**(21), 9911–9917 (2004)
6. D.J. Wales, J.P.K. Doye, Global optimization by basin-hopping and the lowest energy structures of Lennard-Jones clusters containing up to 110 atoms. J. Phys. Chem. A **101**(28), 5111–5116 (1997)
7. R. Martoňák, A. Laio, M. Bernasconi, C. Ceriani, P. Raiteri, F. Zipoli, M. Parrinello, Simulation of structural phase transitions by metadynamics. Zeitschrift für Kristallographie-Crystalline Materials **220**(5-6), 489–498 (2005)
8. S.M. Woodley, P.D. Battle, J.D. Gale, C.A. Catlow, The prediction of inorganic crystal structures using a genetic algorithm and energy minimisation. Phys. Chem. Chem. Phys. **1**(10), 2535–2542 (1999)
9. C.J. Pickard, R.J. Needs, Ab initio random structure searching. J. Phys.: Condens. Matter **23**(5), 053201 (2011)
10. C.W. Glass, A.R. Oganov, N. Hansen, Uspex–evolutionary crystal structure prediction. Comput. Phys. Commun. **175**(11-12), 713–720 (2006)
11. Y. Wang, J. Lv, L. Zhu, Y. Ma, Calypso: a method for crystal structure prediction. Comput. Phys. Commun. **183**(10), 2063–2070 (2012)
12. I.A. Kruglov, A.V. Yanilkin, Y. Propad, A.B. Mazitov, P. Rachitskii, A.R. Oganov, Crystal structure prediction at finite temperatures. npj Comput. Mater. **9**(1), 197 (2023)
13. S. Ono, T. Kikegawa, Y. Ohishi, J. Tsuchiya, Post-aragonite phase transformation in caco3 at 40 gpa. Am. Miner. **90**(4), 667–671 (2005)
14. A.R. Oganov, C.W. Glass, S. Ono, High-pressure phases of caco3: crystal structure prediction and experiment. Earth Planet. Sci. Lett. **241**(1-2), 95–103 (2006)

15. M. Murakami, K. Hirose, K. Kawamura, N. Sata, Y. Ohishi, Post-perovskite phase transition in mgsio3. Science **304**(5672), 855–858 (2004)
16. A.R. Oganov, S. Ono, Theoretical and experimental evidence for a post-perovskite phase of mgsio3 in earth's d layer. Nature **430**(6998), 445–448 (2004)
17. Y. Akahama, M. Nishimura, K. Kinoshita, H. Kawamura, Y. Ohishi, Evidence of a fcc-hcp transition in aluminum at multimegabar pressure. Phys. Rev. Lett. **96**(4), 045505 (2006)
18. J. Lv, Y. Wang, L. Zhu, Y. Ma, Predicted novel high-pressure phases of lithium. Phys. Rev. Lett. **106**(1), 015503 (2011)
19. P. Li, G. Gao, Y. Wang, Y. Ma, Crystal structures and exotic behavior of magnesium under pressure. J. Phys. Chem. C **114**(49), 21745–21749 (2010)
20. L. Zhu, H. Wang, Y. Wang, J. Lv, Y. Ma, Q. Cui, Y. Ma, G. Zou, Substitutional alloy of Bi and Te at high pressure. Phys. Rev. Lett. **106**(14), 145501 (2011)
21. H. Liu, Q. Li, L. Zhu, Y. Ma, Superhard and superconductive polymorphs of diamond-like bc3. Phys. Lett. A **375**(3), 771–774 (2011)
22. H. Liu, Q. Li, L. Zhu, Y. Ma, Superhard polymorphs of diamond-like bc7. Solid State Commun. **151**(9), 716–719 (2011)
23. S.Y. Savrasov, D.Y. Savrasov, Electron-phonon interactions and related physical properties of metals from linear-response theory. Phys. Rev. B **54**, 16487–16501 (1996). (Dec)
24. C. Stampfl, W. Mannstadt, R. Asahi, A.J. Freeman, Electronic structure and physical properties of early transition metal mononitrides: density-functional theory lda, gga, and screened-exchange lda flapw calculations. Phys. Rev. B **63**, 155106 (2001). (Mar)
25. A. Sihi, S.K. Pandey, A detailed electronic structure study of vanadium metal by using different beyond-dft methods. Eur. Phys. J. B **93**, 1–8 (2020)
26. W.G. Aulbur, M. Städele, A. Görling, Exact-exchange-based quasiparticle calculations. Phys. Rev. B **62**, 7121–7132 (2000). (Sep)
27. S.V. Faleev, M. van Schilfgaarde, T. Kotani, All-electron self-consistent gw approximation: application to si, mno, and nio. Phys. Rev. Lett. **93**, 126406 (2004)
28. M. van Schilfgaarde, T. Kotani, S.V. Faleev, Adequacy of approximations in gw theory. Phys. Rev. B **74**, 245125 (2006)
29. M. Shishkin, G. Kresse, Self-consistent gw calculations for semiconductors and insulators. Phys. Rev. B **75**, 235102 (2007)
30. M. Shishkin, M. Marsman, G. Kresse, Accurate quasiparticle spectra from self-consistent gw calculations with vertex corrections. Phys. Rev. Lett. **99**, 246403 (2007)
31. T. Bredow, A.R. Gerson, Effect of exchange and correlation on bulk properties of mgo, nio, and coo. Phys. Rev. B **61**, 5194–5201 (2000)
32. C. Franchini, V. Bayer, R. Podloucky, J. Paier, G. Kresse, Density functional theory study of mno by a hybrid functional approach. Phys. Rev. B **72**, 045132 (2005)
33. J.E. Coulter, E. Manousakis, A. Gali, Limitations of the hybrid functional approach to electronic structure of transition metal oxides. Phys. Rev. B **88**, 041107 (2013)
34. A. Stroppa, G. Kresse, The shortcomings of semi-local and hybrid functionals: what we can learn from surface science studies. New J. Phys. **10**(6), 063020 (2008)
35. F. Tran, S. Ehsan, P. Blaha, Assessment of the gllb-sc potential for solid-state properties and attempts for improvement. Phys. Rev. Mater. **2**, 023802 (2018)
36. F. Tran, P. Blaha, Importance of the kinetic energy density for band gap calculations in solids with density functional theory. J. Phys. Chem. A **121**(17), 3318–3325 (2017)
37. F. Tran, P. Blaha, Accurate band gaps of semiconductors and insulators with a semilocal exchange-correlation potential. Phys. Rev. Lett. **102**, 226401 (2009)
38. W. Li, C.F.J. Walther, A. Kuc, T. Heine, Density functional theory and beyond for band-gap screening: performance for transition-metal oxides and dichalcogenides. J. Chem. Theory Comput. **9**(7), 2950–2958 (2013)
39. W. Kang, M.S. Hybertsen, Quasiparticle and optical properties of rutile and anatase tio 2. Phys. Rev. B– Condens. Matter Mater. Phys. **82**(8), 085203 (2010)
40. H. Tang, H. Berger, P.E. Schmid, F. Levy, G. Burri, Photoluminescence in tio2 anatase single crystals. Solid State Commun. **87**(9), 847–850 (1993)

41. J. Pascual, J. Camassel, H. Mathieu, Resolved quadrupolar transition in ti o 2. Phys. Rev. Lett. **39**(23), 1490 (1977)
42. M. Shishkin, M. Marsman, G. Kresse, Accurate quasiparticle spectra from self-consistent gw calculations with vertex corrections. Phys. Rev. Lett. **99**(24), 246403 (2007)
43. V. Srikant, D.R. Clarke, On the optical band gap of zinc oxide. J. Appl. Phys. **83**(10), 5447–5451 (1998)
44. C. Rödl, F. Fuchs, J. Furthmüller, F. Bechstedt, Quasiparticle band structures of the antiferromagnetic transition-metal oxides mno, feo, coo, and nio. Phys. Rev. B-Condens. Matter Mater. Phys. **79**(23), 235114 (2009)
45. H.K. Bowen, D. Adler, B.H. Auker, Electrical and optical properties of feo. J. Solid State Chem. **12**(3-4), 355–359 (1975)
46. S.V. Faleev, M. Van Schilfgaarde, T. Kotani, All-electron self-consistent gw approximation: application to si, mno, and nio. Phys. Rev. Lett. **93**(12), 126406 (2004)
47. R. Zimmermann, P. Steiner, R. Claessen, F. Reinert, S. Hüfner, P. Blaha, P. Dufek, Electronic structure of 3d-transition-metal oxides: on-site coulomb repulsion versus covalency. J. Phys.: Condens. Matter **11**(7), 1657 (1999)
48. S. Lany, Band-structure calculations for the 3 d transition metal oxides in gw. Phys. Rev. B-Condens. Matter Mater. Phys. **87**(8), 085112 (2013)
49. S.F. Cogan, N.M. Nguyen, S.J. Perrotti, R.D. Rauh, Optical properties of electrochromic vanadium pentoxide. J. Appl. Phys. **66**(3), 1333–1337 (1989)
50. M.B. Johansson, G. Baldissera, I. Valyukh, C. Persson, H. Arwin, G.A. Niklasson, L. Österlund, Electronic and optical properties of nanocrystalline wo3 thin films studied by optical spectroscopy and density functional calculations. J. Phys.: Condens. Matter **25**(20), 205502 (2013)
51. A. Enesca, L. Andronic, A. Duta, S. Manolache, Optical properties and chemical stability of wo3 and tio2 thin films photocatalysts. Rom. J. Inf. Sci. Technol. **10**(3), 269–277 (2007)
52. Y. Qiu, M. Yang, H. Fan, Y. Zuo, Y. Shao, X. Yongjun, X. Yang, S. Yang, Nanowires of α-and β-bi 2 o 3: phase-selective synthesis and application in photocatalysis. CrystEngComm **13**(6), 1843–1850 (2011)
53. H. Jiang, Electronic band structures of molybdenum and tungsten dichalcogenides by the gw approach. J. Phys. Chem. C **116**(14), 7664–7671 (2012)
54. K.K. Kam, B.A. Parkinson, Detailed photocurrent spectroscopy of the semiconducting group vib transition metal dichalcogenides. J. Phys. Chem. **86**(4), 463–467 (1982)
55. A. Schleife, F. Fuchs, C. Rödl, J. Furthmüller, F. Bechstedt, Branch-point energies and band discontinuities of iii-nitrides and iii-/ii-oxides from quasiparticle band-structure calculations. Appl. Phys. Lett. **94**(1) (2009)
56. I. Vurgaftman, J.R. Meyer, Band parameters for nitrogen-containing semiconductors. J. Appl. Phys. **94**(6), 3675–3696 (2003)
57. V.I. Anisimov, M.A. Korotin, E.Z. Kurmaev, Band-structure description of mott insulators (nio, mno, feo, coo). J. Phys.: Condens. Matter **2**(17), 3973 (1990)
58. I.I. Mazin, V.I. Anisimov, Insulating gap in feo: correlations and covalency. Phys. Rev. B **55**, 12822–12825 (1997). (May)
59. D. Koller, F. Tran, P. Blaha, Merits and limits of the modified Becke-Johnson exchange potential. Phys. Rev. B **83**, 195134 (2011). (May)
60. M.A.L. Marques, J. Vidal, M.J.T. Oliveira, L. Reining, S. Botti, Density-based mixing parameter for hybrid functionals. Phys. Rev. B **83**, 035119 (2011)
61. F. Giustino, Electron-phonon interactions from first principles. Rev. Mod. Phys. **89**(1), 015003 (2017)
62. L. Chaput, A. Togo, I. Tanaka, G. Hug, Phonon-phonon interactions in transition metals. Phys. Rev. B-Condens. Matter Mater. Phys. **84**(9), 094302 (2011)
63. S. Huberman, R.A. Duncan, K. Chen, B. Song, V. Chiloyan, Z. Ding, A.A. Maznev, G. Chen, K.A. Nelson, Observation of second sound in graphite at temperatures above 100 k. Science **364**(6438), 375–379 (2019)

64. P. Pandey, V. Pandey, S.K. Pandey, Investigating the effect of electronic correlation on transport properties and phononic states of vanadium. Phys. B: Condens. Matter **669**, 415301 (2023)
65. P. Pandey, S.K. Pandey, Ph-node: a dfpt and finite displacement supercell based python code for searching nodes in topological phononic materials. Comput. Phys. Commun. **303**, 109281 (2024)
66. P. Blaha, K. Schwarz, F. Tran, R. Laskowski, G.K.H. Madsen, L.D. Marks, J. Chem. Phys. **152**(7), 074101 (2020)
67. The Elk Code. http://elk.sourceforge.net/
68. X. Gonze, J.-M. Beuken, R. Caracas, F. Detraux, M. Fuchs, G.-M. Rignanese, L. Sindic, M. Verstraete, G. Zerah, F. Jollet, et al., First-principles computation of material properties: the abinit software project. Comput. Mater. Sci. **25**(3), 478–492 (2002)
69. P. Giannozzi, S. Baroni, N. Bonini, M. Calandra, R. Car, C. Cavazzoni, D. Ceresoli, G.L. Chiarotti, M. Cococcioni, I. Dabo, et al., Quantum espresso: a modular and open-source software project for quantum simulations of materials. J. Phys.: Condens. Matter **21**(39), 395502 (2009)
70. J. Hafner, Ab-initio simulations of materials using vasp: density-functional theory and beyond. J. Comput. Chem. **29**(13), 2044–2078 (2008)
71. M.D. Segall, P.J.D. Lindan, M.J. al Probert, C.J. Pickard, P.J. Hasnip, S.J. Clark, M.C. Payne, First-principles simulation: ideas, illustrations and the castep code. J. Phys.: Condens. Matter **14**(11), 2717 (2002)
72. J.M. Soler, E. Artacho, J.D. Gale, A. García, J. Junquera, P. Ordejón, D.S.ánchez-Portal, The siesta method for ab initio order-n materials simulation. J. Phys.: Condens. Matter **14**(11), 2745 (2002)
73. A.P. Thompson, H.M. Aktulga, R. Berger, D.S. Bolintineanu, W.M. Brown, P.S. Crozier, P.J. In't Veld, A. Kohlmeyer, S.G. Moore, T.D. Nguyen, et al., Lammps-a flexible simulation tool for particle-based materials modeling at the atomic, meso, and continuum scales. Comput. Phys. Commun. **271**, 108171 (2022)
74. A. Togo, L. Chaput, T. Tadano, I. Tanaka, Implementation strategies in phonopy and phono3py. J. Phys.: Condens. Matter **35**(35), 353001 (2023)
75. P. Pandey, S.K. Pandey, Ab initio investigation of the lattice dynamics and thermophysical properties of bcc vanadium and niobium. J. Phys.: Condens. Matter **36**(16), 165602 (2024)
76. S. Baroni, P. Giannozzi, A. Testa, Green's-function approach to linear response in solids. Phys. Rev. Lett. **58**, 1861–1864 (1987). (May)
77. A. Togo, First-principles phonon calculations with phonopy and phono3py. J. Phys. Soc. Jpn. **92**(1), 012001 (2023)
78. A. Togo, L. Chaput, T. Tadano, I. Tanaka, Implementation strategies in phonopy and phono3py. J. Phys.: Condens. Matter **35**(35), 353001 (2023)
79. W. Li, J. Carrete, N.A. Katcho, N. Mingo, Shengbte: a solver of the Boltzmann transport equation for phonons. Comput. Phys. Commun. **185**(6), 1747–1758 (2014)

80. A. Cepellotti, J. Coulter, A. Johansson, N.S. Fedorova, B. Kozinsky, Phoebe: a high-performance framework for solving phonon and electron Boltzmann transport equations. J. Phys.: Mater. **5**(3), 035003 (2022)
81. T. Tadano, Y. Gohda, S. Tsuneyuki, Anharmonic force constants extracted from first-principles molecular dynamics: applications to heat transfer simulations. J. Phys.: Condens. Matter **26**(22), 225402 (2014)
82. J.D. Gale, A.L. Rohl, The general utility lattice program (gulp). Mol. Simul. **29**(5), 291–341 (2003)

Chapter 7
DFT in Surface Science and Catalysis

Mohammad Ammar Syaahiran bin Alim and Roshan Thotagamuge

Abstract Heterogenous catalyst is central to energy conversion, environmental remediation, and sustainable chemical synthesis, where performance is governed by interplay between surface structure, adsorption behavior, and reaction kinetics. Density functional theory (DFT) provides a powerful framework for investigating these processes at the atomic scale through quantitative analysis of surface energies, electronic properties, and reaction mechanism. This chapter presents an integrated approach to DFT-based surface catalysis descriptors, for three interconnected components of surface energy and structure, adsorption process, and catalytic activity and reaction pathways. Key topics include surface modeling, defect formation, adsorption thermodynamics, electronic structure, and transition state calculations. By unifying thermodynamic, kinetics, selectivity, and stability considerations with a consistent theoretical framework, this work offers practical strategies for rational catalyst design. Emerging advances in data-driven methods and multiscaling modeling further highlight the evolving role of DFT in predictive catalyst discovery for sustainable technologies.

Keywords Density functional theory · Surface catalysis · Adsorption energy · Heterogeneous catalyst · Electronic structure

M. A. S. bin Alim (✉)
Centre for Advanced Material and Energy Sciences, Universiti Brunei Darussalam, Gadong, Brunei Darussalam
e-mail: syaahiran.alim@outlook.com

R. Thotagamuge (✉)
Department of Optometry, Faculty of Allied Health Sciences, University of Sri Jayewardenepura, Nugegoda, Sri Lanka
e-mail: roshan@sjp.ac.lk

R. Thotagamuge (ed.), *Density Functional Theory*,
https://doi.org/10.1007/978-981-95-9745-1_7

7.1 Introduction

The study of surfaces and interfaces is central to understanding physical and chemical surface phenomena and reactions, particularly in heterogenous catalysis involving solid–liquid and solid–gas interfaces. Catalytic surfaces provide active sites that enable a cyclic process: adsorption reactants, surface diffusion and dissociation, surface reaction, desorption of products, and diffusion into the surrounding phase [1]. The performance of heterogeneous catalysts is determined by their ability to promote these chemical transformations efficiently, which encompasses four key pillars of thermodynamics, kinetics, selectivity, and stability. Thermodynamics governs the feasibility of the adsorption and reaction steps, kinetics control reaction rates through activation barriers, selectivity determines the preferential of adsorption and desired product formation and stability ensures long-term catalytic performance without deactivation or poisoning the active sites.

In recent decades, density functional theory (DFT) has become a cornerstone of surface science and catalysis, providing an atomistic perspective on surface interactions [2]. DFT enables systematic analysis of surface structures, electronic properties, adsorption energetics, and reaction mechanisms than traditional trial-and-error experimental approaches. It has facilitated the development of descriptor-based approaches from the Sabatier principle, which guide the screening and rational design of catalytic materials.

Despite these advances, numerous theoretical studies continue to focus on isolated descriptors or individual elementary steps, treating surface structure, adsorption, and reaction kinetics as independent phenomena. Such simplified approaches may overlook the complexity of real catalytic system, including the effects of defects, dopants, surface reconstructions, and support interaction that also plays significant roles. Consequently, predictive accuracy and transferability across material classes remain limited.

Recent progress in computational methodology and development has shifted toward more integrated modeling frameworks, such those descriptors are analyzed in unified manner [3, 4]. In this perspective, catalytic performance emerges from coupled effects of thermodynamic, kinetic, selective, and stability-related factors, all of which are ultimately governed by the atomic-scale properties of the surface. DFT provides valuable insights into the surface and adsorption energetics, structural and electronic configurations including its reaction pathways.

In this chapter, a comprehensive framework for DFT-based surface catalysis studies is presented and organized into three interconnected layers:

1. Surface energy and structure, encompassing surface facets, defects, dopants, and electronic descriptors that govern thermodynamic stability and reactivity.
2. Adsorption processes, including adsorption energetics, geometries, interfacial interactions that mediate between catalyst structure and chemical reactivity.
3. Catalytic activity and reaction pathways, addressing electronic structures, electron distributions, reaction steps, transition states, and active barriers.

This integrated perspective provides both conceptual guidance and practical strategy for rational design of efficient, selective, and sustainable catalyst.

7.2 Surface Energy and Structure

The atomic structure and energetic stability of catalyst surfaces constitute the foundation for adsorption and catalytic reactions. Surface morphology, crystallographic orientation, defect distribution, and electronic structure collectively determine the availability and reactivity of active sites. An accurate description of surface energy and structure is therefore essential for understanding catalytic performance in terms of thermodynamics, kinetics, and long-term stability.

7.2.1 Surface Model and Surface Energy

Simulating surface energy through DFT requires an appropriate surface model, commonly a cluster or slab [5]. Cluster models represent small, isolated group of atoms that preservers the symmetry and atomic arrangement of the bulk crystal. Slab models, on the other hand, slice the bulk crystal and impose periodic boundary conditions in two dimensions, with a finite thickness along the surface normal separated by a vacuum layer (typically ~ 20 Å) to prevent interactions between periodic images [6]. The slab is cleaved along planes defined by Miller indices $\{hkl\}$, and its thickness (usually $\geq$ 5 to 10 atomic layers) to ensure bulk-like properties in the interior layers [7, 8].

The slab structure undergoes relaxation to minimize total energy and optimize atomic position. Surface atoms experience unsaturated bonding and forces different from those in the bulk, leading to inward and outward relaxations that reduce surface energy. The surface energy (γ) quantifies the energy required to create the surface from the bulk and is expressed as [9]

$$\gamma = \frac{E_{\mathrm{slab}} - n_{\mathrm{slab}} E_{\mathrm{bulk}}}{2A_{\mathrm{slab}}} \tag{7.1}$$

where E_{slab} and E_{bulk} are the total energies of the slab surface model and the energy per atom of the bulk, n_{slab} is the total number of atoms in the slab model, A_{slab} is the surface area of the slab, and the factor 2 accounts for the two surfaces in slab model. Low surface energy corresponds to greater thermodynamic stability. Surface energy also influences adsorption where surfaces with higher energies can interact more strongly with reactants, although excessive high or low energies may reduce catalytic efficiency [10, 11]. Thermodynamically, low-energy surfaces dominate equilibrium crystal shape (Wulff construction), but under reaction conditions, metastable high-energy surfaces may be catalytically relevant due to kinetic stabilization.

7.2.2 *Influence of Surface Structure on Surface Energy*

Surface structure significantly affects surface energy through atomic arrangement, coordination number, and bonding strength. Crystallographic orientation (low- or high-index facets) determines the density of low-coordinated atoms and surface energy. High-index facets often exhibit higher densities of low-coordinated sites, enhancing adsorption and catalytic activity [12–14]. In some cases such in metal oxides, specific low-index facets (e.g., anatase TiO_2 {110} also display high surface energy due to exposed dangling bonds [15].

Surface reconstruction involves changes in surface periodic or symmetry to stabilize the surface under given conditions [16–19]. For instance, the anatase TiO_2 {110} surface often reconstructs to a (1×4) configuration to minimize surface energy and provide preferential adsorption sites as illustrated in Fig. 7.1.

Catalytic surface model often involved point defects, extended defects, dopant, and deviations from stoichiometry commonly include vacancies and incorporation of substitutional and interstitial impurities which play decisive roles in catalytic behavior. Introduction of defects allow localized electronic states within the band gap or near Fermi level, enhance adsorption strength, and tune surface acidity, basicity, and redox properties. Provide a direct route to modifying both thermodynamic stability and kinetic activity of the surface. Feasibility of the defect formation relates to its formation energy. The formation energy of defect in charge state q is given by [20, 21]

$$E_f^q = E_{\text{defect}} - E_{\text{pristine}} - \sum n_i \mu_i + q(E_{\text{F}} + E_{\text{VBM}}) \tag{7.2}$$

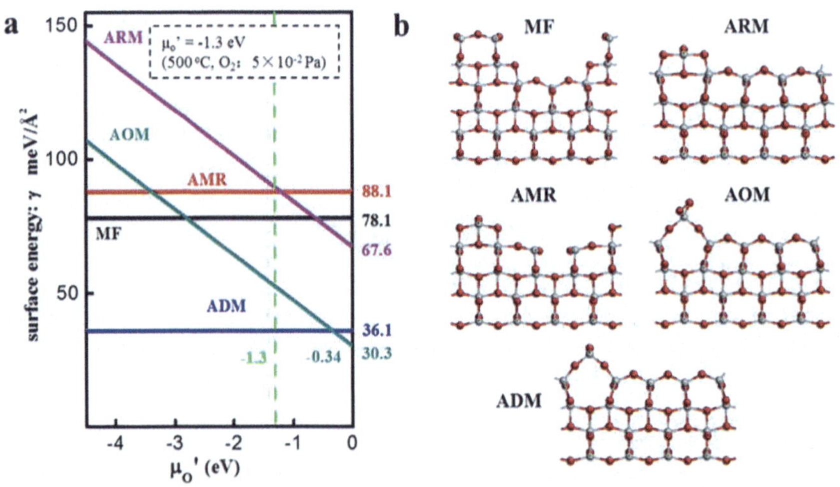

Fig. 7.1 **a** Calculated surface energies vs oxygen chemical potential μ_O' corresponds to **b** surface reconstruction of TiO_2 (1×4) {001} [19]

where E_{defect} and E_{pristine} are the total energies of defect and pristine systems, n_i and μ_i are the number and chemical of missing atoms in the defect systems, E_{F} is the Fermi level, and E_{VBM} is the valence band maximum.

7.3 Adsorption Process

Adsorption represents the primary interaction between reactant molecules and surface dictates chemical reactivity, geometry, and electronic nature properties. Adsorption process undergoes either physisorption and chemisorption [22, 23]. Physisorption is commonly described as the weak adsorption of the reactant on the surface through weak intermolecular forces i.e. van der Waals forces, while chemisorption involves the formation of a chemical bond between the adsorbate and surface atoms. Effective adsorption ensures that reactants are correctly positioned for the reaction. Adsorbate can bind to different surface sites; the top site (on top of a single atom), bridge site (between two atoms), and hollow site (between three or more atoms) as illustrated in Fig. 7.2. The preferred configuration depends on surface coordination, electronic structure, and adsorption symmetry through information on the bond lengths, adsorption distance, and molecular orientations.

The stability of the adsorbates on the surfaces can be calculated through adsorption energy (E_{ads}) as follows [25, 26]

$$E_{\text{ads}} = E_{\text{system}} - E_{\text{surface}} - E_{\text{adsorbate}} \tag{7.3}$$

where E_{system}, E_{surface}, and $E_{\text{adsorbate}}$ are the total energy of the adsorbate–surface complex, the energy of pristine surface, and energy of isolated adsorbate, respectively. A larger negative E_{ads} value indicates a more stable configuration and exothermic reactions.

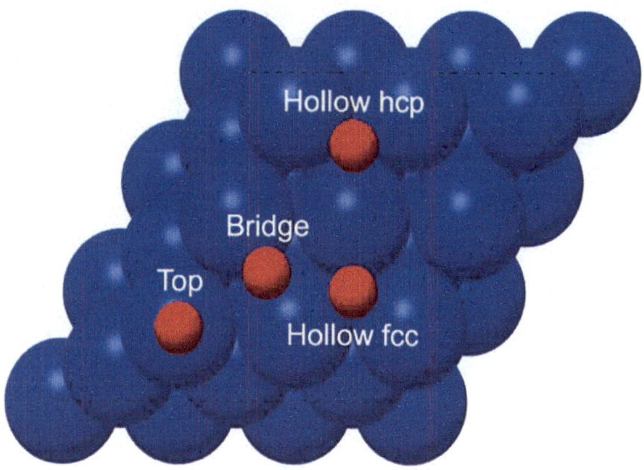

Fig. 7.2 Representation of surface adsorption site on (3 × 3) surface unit of Cu {111} surface [24]

In order, the reactant is to be adsorbed onto the surface, it needs to overcome the activation energy of the reaction. With the presence of a catalyst on the surface enhances the reaction rate through lowering the activation energy. The catalyst acts as an active site for the reaction to occur. Moderate adsorption strength is generally required for optimal catalytic performance, consistent with the Sabatier principle: overly weak adsorption prevent activation, while overly strong adsorption inhibits product desorption. Hence, too strong adsorption can cause catalyst poisoning, while too weak adsorption leads to low activity because reactants do not remain on the surface long enough to react [11]. Thus, a proper balance of adsorption energy leads to efficient catalysis.

Finite-temperature effects are included via Gibbs free energy evaluated as [27–29]

$$\Delta G_{\text{ads}} = \Delta E_{\text{ads}} + \Delta E_{\text{ZPE}} - T\Delta S_{\text{H}} \tag{7.4}$$

where ΔE_{ads} and ΔE_{ZPE} are the differential adsorption energy and the zero-point energy difference between the adsorbed state of the system and the gas phase, T is temperature in Kelvin and ΔS_{H} is the entropy change upon adsorption. This would allow the theoretical prediction correlation experimental conditions such as the "Volcano" plot as represented in Fig. 7.3.

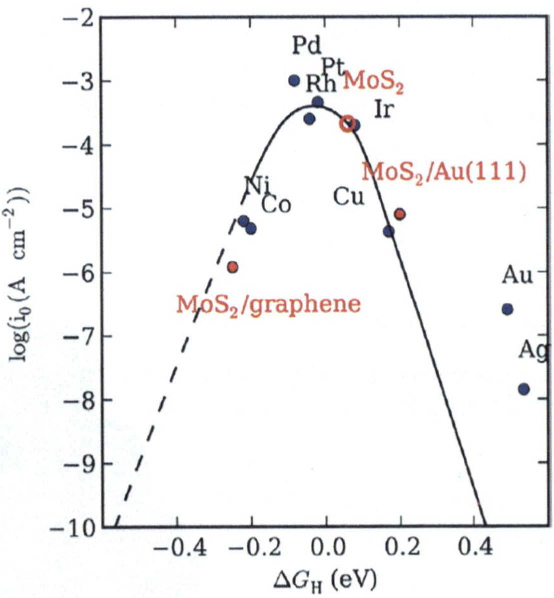

Fig. 7.3 "Volcano" plot for hydrogen evaluation reaction, from Ref. [27]

7.4 Catalytic Activity and Reaction Pathways

Catalytic activity is associated with the ability of a catalyst to accelerate a chemical reaction without being consumed in the process [30]. Predicting catalyst activity is generally complex [31, 32]; however, representative energy-related reaction, such as hydrogen evolution reaction (HER) [27, 29], oxygen evolution reaction (OER) [33], oxygen reduction reaction (OOR) [34], and CO oxidation [35], can serves as benchmark system for evaluating catalytic performance.

The catalytic activity is intrinsically linked to the electronic properties of the catalyst surface. DFT enables detailed analysis of it through quantities such as the distribution of electronic states such as density of states (DOS), projected density of states (PDOS), work function, and band structures, including its electron distribution, transition state, and reaction pathways mechanism, to be investigated.

7.4.1 Electronic Structure and Electron Distribution

Surface electronic structure and electron distribution, as described by *d*-band center theory and frontier molecular orbital (FMO) theory, are critical for understanding catalyst-adsorbate interactions [36]. Adsorbate–metal interactions consist of contributions from *s* and *d* states, while *s* interactions are relatively uniform, trends in adsorption strength are dominated by coupling with *d*-electrons [37]. FMO theory explains that the chemical reaction occurs between molecules through their highest occupied molecular orbital (HOMO) and lowest unoccupied molecular orbital (LUMO) [38]. HOMO and LUMO energies related to the ionization potential (IP $\approx -E_{\mathrm{HOMO}}$) and electron affinities (EA $\approx - E_{\mathrm{LUMO}}$), respectively, and the energy in between indicated as the energy gap E_g [39] which is analogous to the electronic band structure in periodic surface.

In periodic DFT calculations, the surface band structure and DOS enable evaluation of key electronic descriptors, including *d*-band center, valence band maximum, conduction band minimum, and band gap [32, 40]. These parameters are particularly important for semiconductor materials and metal surfaces, as they determine electron availability and reactivity. Furthermore, the electronic structure directly influences the work function (Φ) defined as the minimum energy required to remove an electron from the Fermi level to vacuum and is expressed as [41]

$$\phi = E_{\mathrm{vacuum}} - E_{\mathrm{F}} \tag{7.5}$$

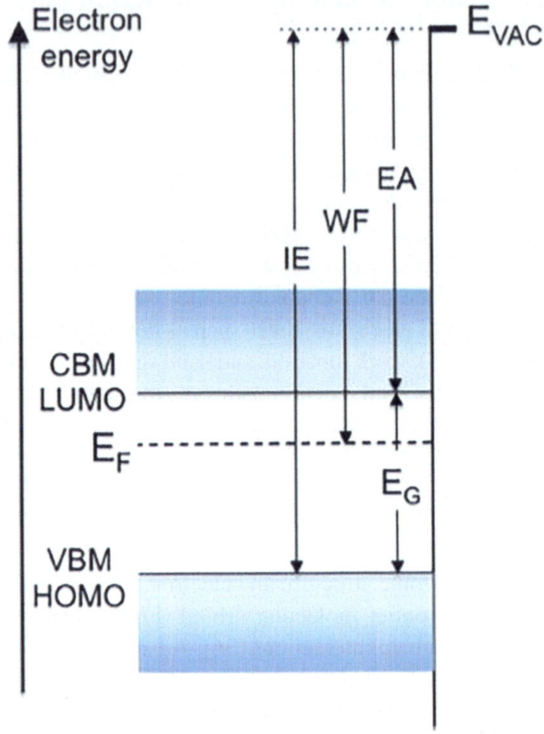

Fig. 7.4 Schematics of semiconductor energy diagram [42]

where E_{vacuum} and E_{F} are the energies with respect to the vacuum potential and Fermi level. The work function is crucial in catalysis because it reflects the surface ability to donate or accept electrons during adsorption and reaction processes. Together, energies of HOMO–LUMO, work function and band structure can be compiled into energy diagram that comprehensively represent the electronic structure, as illustrated in Fig. 7.4.

Since the adsorption involves electronic interactions between surface states and molecular orbitals, these effects can be analyzed using charge density difference (CDD) plot expressed as [43]

$$\Delta\rho = \rho_{\text{system}} - \rho_{\text{surface}} - \rho_{\text{adsorbate}} \tag{7.6}$$

where ρ_{system}, ρ_{surface}, and $\rho_{\text{adsorbate}}$ are electron densities of the combined system, pristine surface, and isolated adsorbate, respectively.

7.4.2 *Transition State and Reaction Pathways*

Transition states (TS) represent the highest-energy structures along reaction paths and are characterized by a single imaginary vibrational frequency along the reaction coordinate. TS structures are located via methods such as nudge elastic band (NEB) [44, 45]. NEB method is widely used to locate minimum energy pathways between the initial (reactant) and final (product) states. This approach creates an intermediate "images" that are connected by elastic springs and optimized to minimize forces perpendicular to the reaction path. Additionally, through climbing-image NEB (CI-NEB) method further refine and estimate the transition states through vibrational frequency analysis. The presence of exactly one imaginary vibrational frequency, ensuring it aligns with the expected reaction pathway. Through connecting the reactants, TS, and products corresponds to their minimum energy path, multi-step reaction pathway network creates energy profile such as the reaction energetics of CO_2 methanation of Ru nanoparticle/TiO_2 catalyst as shown in Fig. 7.5. The activation energy (E_a) is then calculated according to [46]

$$E_a = E_{TS} - E_{reactants} \tag{7.7}$$

where, E_{TS} and $E_{reactants}$ are the energy of the transition state and reactants, respectively. By determining the activation energy from the intermediate energy and the energy barrier, the reaction energetics and kinetics in catalysis reaction can be further analyzed.

7.5 Conclusion and Outlook

DFT has transformed surface catalysis research from an empirical discipline into a predicative, atomistic science by connecting surface structure, adsorption, and reaction pathways to performance. With advances in computation, data-driven methods, and integration with multiscale modeling and operando experiments, it enables rational design of catalysts for thermodynamics, kinetics, selectivity, and stability. Future development of these approaches will accelerate material innovation and support sustainable energy conversion, environmental protection, chemical synthesis, with DFT as central tool for next-generation catalyst discovery.

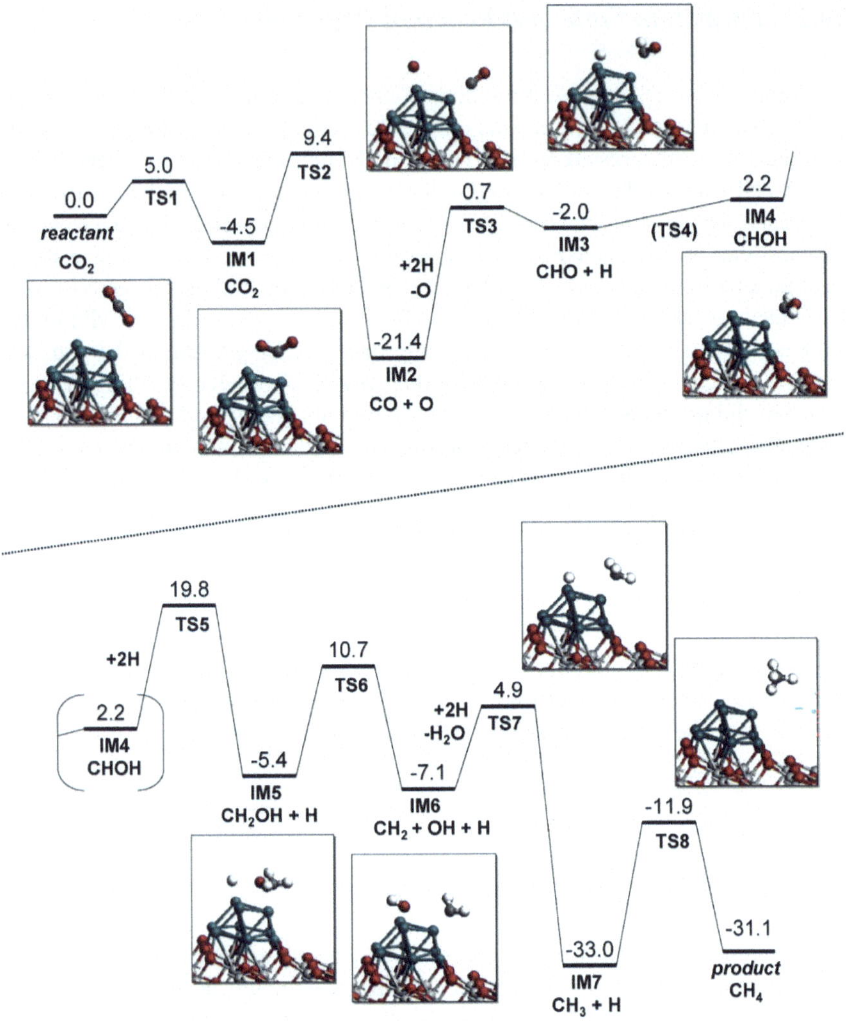

Fig. 7.5 Reaction mechanism for CO_2 methanation on the top of the Ru/TiO_2 catalyst [47]

References

1. M. Bowker, E. Fourré, Appl. Surf. Sci. **254**, 4225 (2008)
2. N. Shan, M. Zhou, M.K. Hanchett, J. Chen, B. Liu, Mol. Simul. **43**, 861 (2017)
3. H. Park, Y. Kim, S. Choi, H.J. Kim, J. Energy Chem. **91**, 645 (2024)
4. X. Zhang, H. Chen, K. Li, B. Wen, J. Ma, Q. Wu, A.C.S. Appl, Nano Mater. **7**, 14288 (2024)
5. D.S. Sholl, J.A. Steckel, *Density Functional Theory: A Practical Introduction* (Wiley, Hoboken, N.J., 2009)
6. S.-H. Yoo, M. Todorova, D. Wickramaratne, L. Weston, C.G.V.D. Walle, J. Neugebauer, NPJ Comput. Mater. **7**, 58 (2021)
7. C.G. Van De Walle, J. Neugebauer, J. Appl. Phys. **95**, 3851 (2004)

8. G. Kresse, O. Dulub, U. Diebold, Phys. Rev. B **68**, 245409 (2003)
9. R. Tran, Z. Xu, B. Radhakrishnan, D. Winston, W. Sun, K.A. Persson, S.P. Ong, Sci. Data **3**, 160080 (2016)
10. H. Zhuang, A.J. Tkalych, E.A. Carter, J. Phys. Chem. C **120**, 23698 (2016)
11. E. Roduner, Chem. Soc. Rev. **43**, 8226 (2014)
12. L. Zhang, W. Niu, G. Xu, Nano Today **7**, 586 (2012)
13. J. Wang, S.-Q. Wang, Surf. Sci. **630**, 216 (2014)
14. C. Xiao, B.-A. Lu, P. Xue, N. Tian, Z.-Y. Zhou, X. Lin, W.-F. Lin, S.-G. Sun, Joule **4**, 2562 (2020)
15. Z. Zhou, Y. Yu, Z. Ding, M. Zuo, C. Jing, Eur. J. Inorg. Chem. **2018**, 683 (2018)
16. F. Polo-Garzon, Z. Bao, X. Zhang, W. Huang, Z. Wu, ACS Catal. **9**, 5692 (2019)
17. L. Jin, P.X. Xu, Y. Zeng, L. Lu, J. Barthel, T. Schulthess, R.E. Dunin-Borkowski, H. Wang, C.L. Jia, Sci. Rep. **7**, 39698 (2017)
18. G. Li, K. Fang, Y. Ou, W. Yuan, H. Yang, Z. Zhang, Y. Wang, Prog. Nat. Sci.: Mater. Int. **31**, 1 (2021)
19. W. Yuan, Y. Wang, H. Li, H. Wu, Z. Zhang, A. Selloni, C. Sun, Nano Lett. **16**, 132 (2016)
20. A. Klein, D. Sudarikov, Phys. Chem. Chem. Phys. **27**, 6390 (2025)
21. S. Zhang, J. Northrup, Phys. Rev. Lett. **67**, 2339 (1991)
22. M. Králik, Chem. Pap. **68**, (2014)
23. H. Swenson, N.P. Stadie, Langmuir **35**, 5409 (2019)
24. E.O. Bartaquim, R.C. Bezerra, A.F.B. Bittencourt, J.L.F. Da Silva, Phys. Chem. Chem. Phys. **24**, 20294 (2022)
25. G. Dai, L. Chen, X. Zhao, J. Mater. Sci. **54**, 1395 (2019)
26. Y. Liu, J. Li, W. Hou, Q. Zhou, W. Zeng, J. Mater. Res. Technol. **18**, 4236 (2022)
27. C. Tsai, F. Abild-Pedersen, J.K. Nørskov, Nano Lett. **14**, 1381 (2014)
28. D. Loffreda, Surf. Sci. **600**, 2103 (2006)
29. T. He, G. Gao, L. Kou, G. Will, A. Du, J. Catal. **354**, 231 (2017)
30. E.B. Maxted, in *Advances in Catalysis* (Elsevier, 1951), pp. 129–178
31. F. Zaccaria, L. Sian, C. Zuccaccia, and A. Macchioni, in *Advances in Organometallic Chemistry* (Elsevier, 2020), pp. 1–78
32. Z.W. Chen, L.X. Chen, Z. Wen, Q. Jiang, Phys. Chem. Chem. Phys. **21**, 23782 (2019)
33. Y. Liu, M. Liao, Y. Hu, T.-G. Lee, R. Koutavarapu, S.G. Peera, C. Liu, Batteries **9**, 175 (2023)
34. M. Reda, H.A. Hansen, T. Vegge, Catal. Today **312**, 118 (2018)
35. H. Jin, H. Zhou, Y. Zhang, Sensors **17**, 1898 (2017)
36. D. Mittal, M. Ahlawat, V. Govind Rao, Adv. Mater. Inter. **9**, 2102383 (2022)
37. B. Hammer, J. K. Nørskov, in *Advances in Catalysis* (Elsevier, 2000), pp. 71–129
38. K. Fukui, T. Yonezawa, H. Shingu, J. Chem. Phys. **20**, 722 (1952)
39. R.G. Pearson, Proc. Natl. Acad. Sci. U.S.A. **83**, 8440 (1986)
40. S. Lu, Z.W. Chen, C. Li, H.H. Li, Y.F. Zhao, Y.Y. Gong, L.Y. Niu, X.J. Liu, T. Wang, C.Q. Sun, J. Mater. Chem. A **4**, 14827 (2016)
41. H.H. Arefi, G. Fagas, J. Phys. Chem. C **118**, 14346 (2014)
42. A. Kahn, Mater. Horiz. **3**, 7 (2016)
43. D. Tian, K. Li, Y. Wei, X. Zhu, C. Zeng, X. Cheng, Y. Zheng, H. Wang, Phys. Chem. Chem. Phys. **20**, 11912 (2018)
44. G. Henkelman, B.P. Uberuaga, H. Jónsson, J. Chem. Phys. **113**, 9901 (2000)
45. G. Henkelman, H. Jónsson, J. Chem. Phys. **113**, 9978 (2000)
46. Z.-P. Liu, P. Hu, A. Alavi, J. Am. Chem. Soc. **124**, 14770 (2002)
47. S. Akamaru, T. Shimazaki, M. Kubo, T. Abe, Appl. Catal. A **470**, 405 (2014)

Chapter 8
DFT on Nanomaterials

Pubudu Samarasekara

Abstract Density functional theory (DFT) has become an indispensable computational tool for understanding and predicting the physicochemical properties of nanomaterials at the atomic and molecular levels. This chapter presents a comprehensive overview of the application of DFT in the study of nanomaterials, highlighting its role in elucidating electronic structure, surface reactivity, optical behavior, and adsorption mechanisms. Emphasis is placed on how DFT enables the rational design and optimization of nanomaterials for applications in sensing, catalysis, energy storage, and biomedicine. The chapter discusses the influence of size, shape, surface functionalization, and defect states on nanomaterial properties, with representative case studies illustrating structure–property relationships. In addition, current challenges, limitations, and emerging trends in DFT-based nanomaterials research are critically examined. By bridging theoretical modeling with experimental observations, this chapter demonstrates how DFT contributes to accelerating nanomaterial development and guiding next-generation nanotechnology innovations.

Keywords Density functional theory (DFT) · Nanomaterials · Electronic structure · Surface interactions · Computational nanotechnology

8.1 Properties of Nanomaterials

Nanomaterials refer to the materials with size ranging from 1 to 100 nm. Due to the higher effective surface area of nanomaterials, absorption and adsorption of nanomaterials are exorbitant. Nanorods, nanowires, nanofibers, nanoparticles, nanolayers, nanotubes are few well-known nanostructures. Because layers consist of nanostructures highly absorb photons, the nanostructures find potential applications in photovoltaic and solar cell applications. Since thin films with nanostructures adsorb vapor or gas molecules, the nanostructures are prime candidates in gas sensor applications. Some other important properties of nanomaterials are quantum effects

P. Samarasekara (✉)
Department of Physics, Faculty of Science, University of Peradeniya, Peradeniya, Sri Lanka
e-mail: pubudus@sci.pdn.ac.lk

R. Thotagamuge (ed.), *Density Functional Theory*,
https://doi.org/10.1007/978-981-95-9745-1_8

and their ability to interact with biological molecules. In addition, nanomaterials are employed in drug delivery, imaging, diagnostics, transistors, sensors, displays, batteries, supercapacitors, fuel cells, water purification, dye removal, air purification, and strengthening materials.

Many different types of nanomaterials can be found in applications. Fullerenes, graphene, multi-walled (MWCNT), and single-walled (SWCNT) carbon nanotubes are the main types of carbon-based nanomaterials. Fullerenes such as C60 are spherical carbon molecules. Graphene is a single layer of carbon atoms arranged in a two-dimensional honeycomb lattice. Metal nanoparticles such as gold, silver, platinum, etc. and quantum dots such as nanoscale semiconductor particles with quantum mechanical properties are the metal-based nanomaterials. Nanostructured ceramics with high hardness, chemical stability and thermal resistance, and metal oxides such as TiO_2, ZnO, CuO, Cu_2O, and SnO are examples for ceramic-based nanomaterials. Clay, carbon nanotubes or graphene nanocomposites, and dendrimers with tree-like structures are the members of polymer-based nanomaterials. Liposomes with a lipid bilayer and solid–lipid nanoparticles (SLNs) are the examples for lipid-based nanomaterials. Nanocellulose derived from plants or bacteria and protein-based nanomaterials derived from virus-like particles are the biological nanomaterials.

Density functional theory (DFT) is one of the most important theoretical tools utilized to investigate the properties of nanomaterials. Both time-dependent density functional theory (TDDFT) and time-independent density functional theories are applied depending on the considered problem. TDDFT is applied to study the dynamic systems with time-dependent potential energies such as electric and magnetic fields. Local density approximation (LDA), generalized gradient approximation (GGA), and meta-GGA are the main approaches employed in DFT. Hubbard-U correction has been introduced to overcome the shortcomings of LDA and GGA. Among the advantages of the Hubbard correction, it improves the treatment of localized d and f electrons, in which DFT itself fails to delineate accurately due to the self-interaction error. In addition, it provides better band gap corrections, corrects magnetic properties, enhanced structural properties, accurate localized states. Disadvantages of the Hubbard-U correction are its empirical nature, computational cost, limited transferability, complex implementation, and potential overcorrection.

Computer software packages such as ABINIT, CASTEP, CPMD, VASP, Quantum Espresso, CP2K, and OCTOPUS are widely utilized to devolve into nanomaterials. Among these software packages, Quantum Espresso and ABINIT are open-source with broad capabilities. CASTEP and VASP offer high accuracy and performance for solid-state materials. CPMD is vastly applied for molecular dynamic with DFT. CP2K is efficient for large-scale simulations and mixed systems. OCTOPUS handles time-dependent simulations and electron dynamics. Studying the variation of structural, electrical, and magnetic properties with the positions of different atoms leads to synthesize new nanostructures with enhanced properties.

Considering the nanostructures, DFT is employed to determine (i) the electronic structure such as band structure, density of states (DOSs), and Femi level; (ii) the structural properties such as equilibrium geometry, lattice parameters, bond lengths, and bond angles; (iii) the mechanical properties such as elastic constants, bulk modulus, shear modulus, and Young's modulus; (iv) the optical properties such as dielectric function and refractive index; (v) the magnetic properties such as magnetic moments and spin density; (vi) the chemical properties such as adsorption energies, reaction energies, and charge density distribution. TDDFT is applied to investigate (i) the optical properties such as absorption spectra, excitation energies, and optical transitions; (ii) the dynamic response properties such as dielectric function and plasmon resonances; (iii) the excited state properties such as excited state lifetimes, singlet and triplet excitations; (iv) the charge transfer dynamics such as non-adiabatic transitions and photoinduced processes.

Despite its success and versatility, several types of errors and limitations can arise in DFT calculations, particularly when dealing with nanomaterials. The choice of exchange–correlation (XC) functional significantly affects the accuracy of DFT calculations. Commonly used functionals like local density approximation (LDA) and generalized gradient approximation (GGA) may not accurately capture van der Waals interactions or strong correlation effects, which are important in nanomaterials. Hybrid functionals, which mix a portion of exact exchange from Hartree–Fock theory with DFT, can provide better accuracy but are computationally more expensive. Nanomaterials often have properties that depend on their size and shape. In DFT calculations, using finite clusters or supercells can lead to finite size effects, where the calculated properties may not accurately represent those of the infinite or bulk system. The choice of basis set, whether plane waves, localized atomic orbitals, or others, can introduce errors. An incomplete basis set may lead to inaccurate results, while a very large basis set increases computational cost. The use of pseudopotentials to replace the core electrons can introduce errors if they are not carefully constructed or transferred between different chemical environments. Achieving self-consistent convergence in DFT calculations can be challenging, especially for systems with a large number of atoms or complex electronic structures. Poor convergence can lead to inaccurate results. For nanomaterials with magnetic properties, incorrect treatment of spin states or magnetic ordering can result in significant errors. The choice of initial magnetic moments and the treatment of spin–orbit coupling are crucial. In nanomaterials, surface and interface effects are prominent. Standard DFT methods may not always accurately capture these effects, especially when dealing with surface reconstructions or adsorbates. Van der Waals interactions are crucial in many nanomaterials, such as those involving layered structures like graphene or MoS2. Standard DFT functionals often fail to capture these interactions accurately, necessitating the use of specialized dispersion-corrected functionals (e.g., DFT-D). DFT is typically performed at zero temperature and does not inherently include entropic effects. For nanomaterials where thermal effects are significant, additional methods (such as molecular dynamics or phonon calculations) are required to capture temperature-dependent properties. DFT in its standard form is a ground-state theory and may not accurately describe charge transfer processes or excited states. Time-dependent DFT (TDDFT) or many-body perturbation theory (e.g., GW approximation) might be needed for such cases. Understanding these limitations and carefully choosing appropriate methods and parameters can mitigate some of these errors.

Additionally, comparing DFT results with experimental data and higher-level computational methods (like coupled-cluster or quantum Monte Carlo) can help validate the findings.

DFT and TDDFT simulations have been performed for nano-sized Cu(II) complexes fabricated from the reflux and ball mailing techniques using Schiff base. The optimized molecular structures and atomic numbering schemes for the ligand and its Cu(II) complexes have been obtained. Bond lengths and angles have been compared between the ligand and its Cu(II) complexes, revealing changes due to complex formation. For instance, bond lengths like C(12)-N(11) and C(9)-O(26) increased after complex formation, while others like C(8)-C(9) decreased. The [Cu(HL)Cl] complex was found to have a square planar geometry, confirmed by bond angles and a calculated τ_4 value close to zero. The $[CuH_2LCl_2(H_2O)_2]$ complex exhibited an octahedral geometry. Mulliken atomic charge calculations indicated a transfer of negative charge from the ligand to the Cu(II) ions upon complexation. The coordination sites of ligand and geometrics for obtained Cu(II) complexes depend on the preparation method of the samples. TDDFT has been used to calculate electronic excitation energies with HOMO/LUMO levels, yielding UV–visible spectra that matched experimental results. In the [Cu(HL)Cl] complex, transitions like HOMO-3 $\rightarrow$ LUMO + 1 and HOMO-2 $\rightarrow$ HOMO were identified, while for the $[CuH_2LCl_2(H_2O)_2]$ complex, transitions such as HOMO-3 $\rightarrow$ LUMO and HOMO-4 $\rightarrow$ LUMO + 1 were observed. MEP maps were generated to visualize the charge distribution and potential reactive sites in the molecules. For the ligand H2L, negative potentials were found around the OH and carbonyl groups, suggesting these as coordination sites with Cu(II). After complexation, changes in the MEP maps indicated electron transfer from the ligand to the Cu(II) ions, consistent with the coordination chemistry [1].

DFT studies of MgO nanosheets with (111) facet synthesized from wet chemical have been performed. Gold nanoparticles have been used as catalyst. Charge densities were calculated for MgO substrates with oxygen atoms on the top layer: (001), (110), and O-terminated (111). The results agree with some previous DFT studies of MgO surfaces. On (001) and (110) surfaces, the charges of O and Mg are close to their bulk values. The (111) surfaces show significant deviations, indicating higher reactivity, especially with O atoms having fewer electrons and Mg atoms having more electrons than in the bulk state. Au_{10} clusters were relaxed on each surface, resulting in various structural deformations due to strain mismatch. The average Au–Au bond length in a relaxed cluster is 2.74 Å. MgO(001) and MgO(110) surfaces cause significant deformations and partial decoupling of Au clusters due to atomic spacing mismatches. On MgO(111) surfaces, smaller deformations occur, and clusters remain on the substrate. Strong binding and structural integrity of the Au cluster are observed on (111) surfaces, while partial debonding and deformations occur on (001) and (110) surfaces. Charge transfer between Au clusters and MgO substrates has been analyzed using Bader charge analysis. On (111)-O and (111)-Mg surfaces, substantial charge transfer occurs, with the Au cluster losing 3.95 e to (111)-O and gaining 7.15 e from (111)-Mg. This significant charge transfer is mainly localized in the six Au atoms at the interface. The MgO surface orientation significantly affects the catalytic activity of gold nanoparticles. Au/MgO(111) systems show higher catalytic activity, possibly due to electron donation to the O-terminated surface, increasing the reactivity of exposed oxygen atoms. Weak binding and charge transfer on (001) and (110) surfaces correlate with lower catalytic activity. DFT results highlight the importance of MgO surface orientation for the binding and charge transfer properties of gold clusters. The strong binding and significant charge transfer on (111) surfaces enhance the catalytic activity of Au/MgO(111) systems. The charge transfer trends are consistent with the tendency of surface atoms to regain their bulk charge states [2].

New complexes of Zn(II), Cu(II), Ni(II), and Co(II) in nano scale have been prepared for antibacterial applications. Zn(II) has been found as a promising material for antibacterial applications. The article details DFT calculations performed on a series of metal nano-complexes derived from thiophene-2-carbaldehyde. The computational work was carried out using the B3LYP hybrid density functional with the 6-311G(d,p) basis set, implemented in the Gaussian 09 package. The structures of the ligand and its metal complexes (with Zn(II), Cu(II), Ni(II), and Co(II)) were fully optimized without symmetry constraints. All structures were optimized, and the results were visualized using GaussView 5.0.9 and Chemcraft 1.6 software. The total energy, HOMO, LUMO, energy gap, and dipole moment were calculated. The energy gap, which indicates chemical reactivity, was found to be the smallest for the Zn(II) complex, suggesting it is the most reactive. The Zn(II) complex had the highest dipole moment, indicating significant polarity. The Zn(II) complex had a relatively low ionization energy and high electron affinity, suggesting a strong ability to donate and accept electrons, respectively. The Zn(II) complex had the highest net charge transfer from the ligand to the metal, indicating a strong interaction between the ligand and the metal ion. The antimicrobial activities were tested against various bacteria and fungi. The free ligand (H2L) showed no significant activity, whereas the metal complexes exhibited varying degrees of antimicrobial properties. The Zn(II) complex demonstrated the highest antimicrobial activity, particularly against Gram-positive and Gram-negative bacteria and fungi. The high activity was attributed to its small crystal size and high reactivity, which enhances penetration through microbial cell membranes. The study correlated the biological activity of the complexes with their calculated ground-state properties. The order of reactivity (based on the energy gap and charge transfer) was Ni(II) > Zn(II) > Co(II) > Cu(II), which matched the observed antimicrobial effectiveness. The high reactivity and polarity of the Zn(II) complex made it particularly effective as an antimicrobial agent. The theoretical findings were consistent with the experimental results, providing a comprehensive understanding of the relationship between the structural properties and biological activity of the metal complexes [3].

Monodispersed PVP/gold and silver nanoparticles have been investigated using combined DFT/FTIR structural studies. Becke's three-parameter hybrid functional (B3LYP) and the electron core potential basis set LANL2DZ together with Gaussian 03 software were employed. For Ag and Au, the LANL2DZ basis set simulates 28 and 60 core electrons, respectively, while the remaining electrons are described by all electron basis sets. The molecular structures of PVP with Ag and Au nanoparticles were optimized. Harmonic frequency calculations were performed to characterize stationary points and identify Fourier transform infrared (FTIR) vibrational modes. Various interaction modes between PVP and metal nanoparticles were proposed via carbonyl group's nonbonding electrons, nitrogen atom, or both. Theoretical FTIR results indicated that the interaction with oxygen atoms of the pyrrolidone ring was the most stable and matched experimental data. UV–Vis spectra showed characteristic absorption peaks for both PVP/Au and PVP/Ag films, attributed to surface plasmon resonance. The optical band gap was calculated from the absorption spectra, with results aligning with experimental data. The particle sizes observed in TEM were consistent with DFT calculations and other experimental results. The study demonstrated that PVP doped with Ag and Au nanoparticles maintained the characteristic polymeric network while interacting predominantly through the carbonyl group's non-bonding electrons. Combined DFT and FTIR analyses provided detailed insights into the molecular interactions and structural properties of the doped polymer matrices [4].

DFT work of three novel N4 asymmetric Schiff bases and their nano-sized cadmium complexes with formula $[Cd^{II}{}_2L^nCl_4]$, ($L = C_{45}H_{40}N_5X$, $X = CH_3$, Cl, OH) synthesized by sonochemical method for efficient, green, low cost, recyclable, and convertable nanocatalytic systems has been described. Calculations were performed using Gaussian 09 with the B3LYP functional. The 6-311G + (d,p) basis set was used for ligand atoms, and the LanL2DZ basis set was used for cadmium atoms. All geometries were optimized without symmetry constraints. Frequency calculations were conducted to verify the nature of the stationary points. ^1H and ^{13}C NMR shifts were computed using the GIAO method and compared with experimental data. The optimized structures of cadmium complexes were obtained and compared with experimental data. Bond lengths and angles showed good agreement with experimental values, including X-ray crystal data. Specific bond lengths and angles of interest included Cd–N, Cd-Cl, C1 = C2 (aromatic) and various C-H and C–C bonds. The energy gap ($\Delta E = E_{LUMO}$—E_{HOMO}) was calculated to assess molecular reactivity. Chemical descriptors such as hardness (η), electronegativity (χ), softness (S), and electrophilicity index (ω) were derived. For UV–Vis spectra, electronic spectra were used to investigate the geometry and electronic transitions of the complexes. IR spectra confirmed the formation of Schiff bases and their metal complexes through characteristic vibrational bands. ^1H and ^{13}C NMR spectra provided insights into the chemical environment of protons and carbons in the complexes. The study effectively demonstrates the use of DFT calculations to predict and analyze the structural, electronic, and spectroscopic properties of cadmium complexes. The computational results showed good agreement with experimental data, validating the theoretical approach used. The formation and characterization of CdO nanoparticles from the thermal decomposition of cadmium complexes were also successfully achieved [5].

The electronic transport properties of CuS (Copper Sulfide) nanostructures, which are important for molecular electronics due to their distinct behavior compared to bulk materials, have been investigated using DFT. The study focuses on a molecular device consisting of three regions: left electrode, right electrode, and the device molecular region. The investigation emphasizes the significance of electrodes and the device region in determining device performance, with a particular interest in inorganic molecules like CuS. The research employs the non-equilibrium Green's function (NEGF) method combined with density functional theory (DFT) to analyze CuS molecular devices. Additionally, the study examines the impact of acetylene and nitroso functional groups adsorbed on CuS nanowires. Key computational tools include the Gaussian 09 package for DFT calculations and the TRANSIESTA in SIESTA package for transport properties analysis. CuS nanowires were optimized using the LANL2DZ basis set with the B3LYP energy functional. The average bond length between Cu and S atoms was found to be 2.07 Å, matching experimental data. HOMO and LUMO values were −4.36 eV and −3.47 eV, respectively, resulting in a band gap of 0.89 eV. The band gap increased with acetaldehyde adsorption and decreased with nitroso functional group adsorption due to changes in orbital interactions. The DOS analysis showed more energy states in the conduction band, indicative of p-type behavior. The transmission spectrum revealed a strong coupling between electrodes and the device at specific energy levels, confirmed by PDOS spectra. Ionization potential (IP) and electron affinity (EA) were used to derive chemical potential, hardness, and electro negativity. CuS nanostructures exhibited a chemical potential of −3.915 eV, hardness of −0.445 eV, and electronegativity of 3.915 eV. Transport properties were studied using a two-probe system under varying bias voltages. The analysis indicated significant electron transmission at zero bias due to the chemical potential interaction between gold electrodes and CuS atoms. The study highlights the importance of theoretical analysis of molecular device regions before fabrication. CuS nanowires, with their unique electronic properties influenced by adsorbed functional groups, demonstrate potential for various applications in molecular electronics. The findings underscore the intricate interplay between structural optimization, electronic properties, and transport behavior in designing effective molecular devices [6].

DFT has been employed to investigate the properties of spherical Fe_3O_4 nanoparticles synthesized from the thermal decomposition of iron (III) nanostructure. The article investigates the vibrational properties, structural characteristics, and interaction between a Schiff base ligand and iron-ion complexes using DFT. Three types of Fe(III) complexes were considered, with a low-spin configuration based on prior studies. The complexes were optimized using the B3LYP/6–31 + G(d) level of theory with the Gaussian 03 package. The optimized structures, vibrational frequencies, molecular electrostatic potential (MEP), and HOMO–LUMO energy gaps were calculated. HOMO–LUMO gaps provide insights into the stability and chemical activity of the complexes. The complexes with coordination numbers 5 and 6 showed HOMO orbitals distributed over iron and nitrate groups, while LUMO orbitals were on the Schiff base. Tetrahedral complexes had HOMO and LUMO orbitals spread across the entire molecule, indicating nitrate anion binding affects charge transfer. MEP maps showed regions of electrophilic (positive) and nucleophile (negative) reactivity, helping identify reactive sites for electrophilic and nucleophile attacks. Natural Bond Orbital (NBO) analysis showed charge transfer mainly from the Schiff base ligand and nitrate groups to the d-orbital of iron, with interaction energies ranging from 1.06 to 28.08 kcal/mol [7].

The adsorption properties of explosive materials such as RDX, TATP, HMTD, TNT, HMX, and PETN with boron nitride (BN) and graphene (G) flakes have been studies using DFT and TDDFT. Circular flakes of BN and graphene with hydrogen-saturated edges were used. The geometries of explosives and their complexes were optimized using B3LYP-D with 6–31+ G(d,p) and 6–311++ G(2df,2pd) basis sets. Binding energies (BEs) were calculated with basis set superposition error (BSSE) correction. Natural Bond Orbital (NBO) analysis and time-dependent DFT (TDDFT) were conducted for electronic and optical properties. Periodic calculations were performed using the Vienna ab initio simulation package (VASP). TNT forms the most stable complexes due to strong π-π interactions with BN and graphene. The calculated binding energies indicate stronger interactions of explosives with BN flakes compared to graphene. Hydrogen atoms in HMTD and TATP orient toward the BN ring, indicative of CH-π interactions. Oxygen atoms in HMX, PETN, and RDX show favorable interactions with boron atoms in BN due to electronegativity differences. Cation-π interactions and CH-π interactions are significant in stabilizing PETN and other complexes. BN-explosive complexes have higher binding energies compared to graphene-explosive complexes, confirming stronger interactions. TNT shows the highest binding energy with both BN (82.00 kJ/mol) and graphene (72.35 kJ/mol). BN-explosive complexes exhibit more significant charge transfer than graphene-explosive complexes, supporting the observed higher binding energies. The TNT–BN complex shows the highest charge transfer due to strong π-π interactions. Adsorption of explosives reduces the HOMO–LUMO gap of BN flakes significantly, especially with TNT. The electronic structure analysis shows contributions to the HOMO and LUMO from both BN/graphene and explosives, depending on the molecule. BN flakes show strong UV absorption around 192 nm. Explosive molecules exhibit absorption peaks corresponding well with experimental values. The absorption spectra of BN-explosive complexes show distinct features, with some complexes displaying quenched absorption due to lack of available unoccupied orbitals. Both BN and graphene complexes are stabilized mainly by dispersion interactions, with additional contributions from electrostatic and orbital interactions. BN-explosive complexes have higher dispersion and electrostatic contributions compared to graphene complexes, particularly for TNT due to B-O interactions. BN flakes exhibit stronger interactions and higher binding energies with explosive molecules compared to graphene, making BN a potentially better material for sensing applications involving explosives [8].

DFT has been utilized to investigate the electronic properties of nano-graphene with hydrogen-terminated edges, which differ significantly from infinite graphene sheets. The study focuses on the zig-zag (trans) and arm-chair (cis) edges of 2D graphene sheets, which are neutralized by hydrogen atoms. Experimental techniques like atomic force microscopy (AFM) and scanning tunneling microscopy (STM) have shown varying electronic properties at these edges. The research employs density functional theory (DFT) with the B3LYP functional to understand these properties theoretically. Three models of graphene sheets with varying sizes and edge structures were generated: a small model (pyrene) with 16 carbon and 10 hydrogen atoms, an intermediate model with 32 carbon and 14 hydrogen atoms, and a large model with

73 carbon and 21 hydrogen atoms. Additionally, a model with all arm-chair edges and a model with a central vacancy were created. The calculations were performed using GAMESS-US software. The zig-zag edges showed higher electron density compared to arm-chair edges. This observation aligns with experimental results from AFM and STM studies. The HOMO–LUMO gap decreases as the size of the graphene sheet increases, suggesting that larger graphene sheets approach metallic behavior. A mono-vacancy in the graphene sheet leads to high electron density around the vacancy and structural deformation. This reduced vacancy is similar to an H3 complex and can induce magnetism, which is consistent with experimental observations of magnetism in hydrogen-irradiated graphene. The electron density varies with the separation between graphene layers. Closer layers (3.4 Å) show less electron population compared to more separated layers (10 Å), indicating that increased separation enhances electrical conductivity. In conclusion, DFT with the B3LYP functional effectively models the electronic properties of nano-graphene, matching experimental findings. The study highlights the importance of edge structure, size, and defects in determining the electronic characteristics of grapheme [9].

The crystal structure of template synthesized $HgLI_2$ has been explained by DFT. The study involves the synthesis and characterization of a new mercury(II) complex with the ligand N^1–(4–nitrobenzylidene)–N^2–(2–((E)–(4–nitrobenzylidene)amino)ethyl)ethane-1,2–diamine. The complex crystallizes in a triclinic system with space group P1 and exhibits a distorted trigonal bipyramidal geometry around the mercury(II) center. The structure includes two iodine atoms and one nitrogen atom from the Schiff base ligand in the equatorial plane, with two iminic nitrogen atoms in the axial positions. The crystal structure reveals two five-membered rings containing mercury, as well as various intermolecular interactions that contribute to a supramolecular network in the solid state. Additionally, DFT study was conducted using the B3LYP/LanL2DZ level of theory to compare theoretical structural and spectral data with experimental results. The DFT calculations included geometry optimization, evaluation of total energy levels of HOMO and LUMO orbitals, molecular electrostatic potential, Mullikan atomic charges, thermodynamic properties, and polarizability properties. The theoretical data showed good agreement with experimental data, including bond lengths, bond angles, and NMR chemical shifts. A small HOMO–LUMO energy gap indicating easier electron excitation and charge transfer interactions within the complex. The HOMO energy levels vary from −19.19271 to −0.23519 a.u., while the LUMO energy levels change from −0.12830 to 12.01581 a.u. The HOMO–LUMO energy gap is low for the titled complex (0.10689 a.u.). Significant electron transfer was observed from ligands to the mercury ion. Stability indicated by low total energy and negative energies of HOMO, LUMO, and neighboring orbitals [10].

DFT has been employed to investigate properties of CuO nanoparticles prepared using thermal decomposition. These CuO nanoparticles have exhibited bactericidal effect against some bacteria. The article discusses the synthesis, characterization, and analysis of a nano-sized copper (II) complex, [Cu(L)], formed through the reaction of 2–((E)–(2–(E–5–bromo–2–hydroxybenzenylideneamino)methyl)–4–bromophenol [H2L] with copper(II) acetate. The complex was studied using both experimental methods and DFT. The geometry of the [Cu(L)] complex was optimized using DFT at the B3LYP/6–31(d) level of theory. The calculated bond lengths and angles were found to be in good agreement with experimental XRD data. Small deviations between experimental and theoretical data were noted, attributed to differences between solid-state and gas-phase conditions. HOMO and LUMO distributions were calculated, showing their presence over the conjugated plane. Weak π-π interactions were noted with distances between centroids of phenyl rings around 4.305 Å. Shift in C = N stretching frequency upon complexation and new bands indicating metal–ligand coordination were observed in FTIR spectrums. FESEM analysis showed semi-spherical shapes for the [Cu(L)] complex with particle sizes around 40–50 nm; post-decomposition, CuO particles were 20–30 nm [11].

Covalent functionalization of a $B_{12}N_{12}$ nano-cage with CO_2 was investigated using DFT. CO_2 can weakly interact with $B_{12}N_{12}$ through physisorption in two stable configurations, with minimal energy release and negligible changes in electronic properties. A $B_{12}N_{12}$ nano-cage is formed by eight 6-membered rings (6-MR) and six 4-membered rings (4-MR). The bonds shared by two 6-MRs have been defined as B_{66}, and the bond shared between a 4-MR and a 6–MR has been defined as B_{64}. CO_2 prefers a [2 + 2] cycloaddition on B-N bonds shared between six-membered and four-membered rings, with higher energy release for adsorption on B_{64} bonds compared to B_{66} bonds. This results in significant geometric distortions and electronic property changes. Physisorption does not alter the electronic properties of $B_{12}N_{12}$, while covalent functionalization slightly reduces the HOMO–LUMO gap, making the material more reactive. Adsorption of a second CO_2 molecule on a B_{64} bond releases slightly more energy, further altering the electronic properties. DFT calculations were performed using B3LYP/6-31G(d) and Møller–Plesset perturbation theory (MP2) methods, confirming that the covalent functionalization has a more significant impact on the material's properties compared to physisorption. The study concludes that covalent functionalization with CO_2 could modify the $B_{12}N_{12}$ properties for various applications, including CO_2 capture and storage [12].

The effect of Hafnium (Hf) and Boron (B) additions on the phase stability, microstructure, and magnetic properties of metastable SmCo7 (1:7H) ribbons using experimental measurements and first-principle DFT calculations has been studied. DFT calculations confirmed the phase stability of the $(Sm_{0.12}Co_{0.88})_{95}Hf_{5-x}B_x$ ribbons. The total energy calculations indicated that the stability of the $SmCo_7$ phase increases with Hf addition up to a certain limit but decreases beyond that point. The disordered $SmCo_7$ structure was modeled by substituting Sm sites with Co–Co dumbbell pairs. Hf atoms were introduced at specific sites to stabilize the structure, while B atoms were added for grain refinement. The Vienna ab initio Simulation Package (VASP) was used for structure optimization and electronic structure calculations, employing the generalized gradient approximation (GGA) with Burke–Ernzerhof (BE) functional. The GGA + U method was applied to correct for strong on-site Coulomb repulsions. The formation energies of different alloy compositions were calculated, revealing that energy values become more negative with increasing Hf addition up to a point. This indicates a more stable structure, but stability decreases with excessive Hf. The study calculated magnetic anisotropy energies and compared them with experimental coercivity values. Saturation magnetization was also computed, showing an increase with higher B concentration. The study provides insights into how Hf and B additions affect the phase stability, microstructure, and magnetic properties of $SmCo_7$ ribbons, with DFT calculations supporting the experimental results [13].

8.2 Quantum Dots and Nanowires

Quantum dots and nanowires are nanomaterials with unique properties that make them suitable for a wide range of applications across various fields. Quantum dots are employed in optoelectronics, biomedical applications, quantum computing, and sensors. Quantum dots are used in LED displays and lighting to produce highly pure colors and enhance energy efficiency. Quantum dots improve the efficiency of photodetectors and solar cells by enhancing light absorption and conversion efficiency. Quantum dots are used as fluorescent markers for imaging and tracking cells and molecules in biological research. They can be used to deliver drugs to specific cells or tissues, providing targeted therapy. Quantum dots are utilized as qubits in quantum computers, potentially leading to more powerful and efficient computing systems. They can be used to detect the presence of specific chemicals or biological molecules due to their sensitivity to changes in the environment.

Nanowires find potential applications in electronics and photonics, energy conversion and storage, sensors, nano-electronics, and thermo-electrics. Nanowires can be used to create transistors at the nanoscale, leading to faster and more efficient electronic devices. They are used in photodetectors for high sensitivity and fast response times. Nanowires can improve the efficiency of solar cells by providing better light absorption and charge separation. They enhance the performance of batteries and supercapacitors by providing a higher surface area for charge storage. Nanowires can detect gases at very low concentrations, making them useful for environmental monitoring and safety applications. They can be used to detect specific biological molecules, making them useful for medical diagnostics and research. They can be employed as interconnects in nanoelectronic circuits, providing high conductivity and small dimensions. Nanowires can enhance the efficiency of thermoelectric materials, which convert heat into electricity, making them useful for energy harvesting applications.

A comprehensive DFT and TDDFT analysis of graphene quantum dots (GQDs) and their functionalized variants using the Gaussian 09 software have been performed. The computations employed various functionals including PBE, B3LYP, PBE0, TPSSh, CAM-B3LYP, and LC-wPBE with the 6-31G(d) basis set, evaluating solvent effects through the Polarizable Continuum Model (PCM). Ground-state geometries (S0) of GQDs were optimized and verified for no imaginary frequencies. Functional groups were replaced by hydrogen atoms for simplification. Absorption wavelengths were computed using TDDFT at optimized geometries, considering the lowest ten singlet excited states. Solvent effects on absorption were evaluated using the non-equilibrium solvation model. PBE and TPSSh functionals underestimated excitation energies, predicting longer wavelengths than experimental data. PBE0, CAM-B3LYP, and LC-wPBE overestimated excitation energies, predicting shorter wavelengths. B3LYP functional provided results closest to experimental values for both absorption and fluorescence wavelengths. Vertical excitation energies for some carbon atoms were calculated in gas phase and solvents (toluene, dichloromethane, water). Absorption maxima were found to redshift in solvents compared to gas phase. Detailed analysis of molecular orbital transitions showed significant contributions from frontier molecular orbitals (HOMO-1, HOMO, LUMO, LUMO+1). Functionalized GQDs exhibited small redshifts in absorption maxima, with the degree of redshift depending on the number and type of substituent groups (e.g., OH, COOH). Fluorescence wavelengths were computed considering solvent relaxation effects. The fluorescence of some carbon atoms in toluene was accurately predicted by the B3LYP functional, agreeing with experimental data and confirming Kasha's rule. Electron transition contributions were analyzed, showing major contributions from specific orbital transitions (LUMO to HOMO-1 and LUMO + 1 to HOMO). Real-space analysis of electronic transitions highlighted primary electron transition sites within GQDs. Functionalized GQDs showed different electron transition patterns compared to non-functionalized GQDs. The study highlights the suitability of B3LYP for predicting electronic excitations in GQDs, demonstrating its accuracy in reproducing experimental absorption and fluorescence spectra. It also emphasizes the impact of functional groups and solvent environment on the optical properties of GQDs [14].

In addition, DFT has been applied to investigate epoxy/graphene quantum dots [15], InGaAs/GaAs quantum dots [16], CdS quantum dots [17], ZnS quantum dots [18], CdTe quantum dots [19], and CdSe quantum dots [20].

DFT has been employed for first-principles calculations on anatase TiO_2 nanowires (NWs) to study their structural and electronic properties based on size, surface coverage, and morphology. Increasing the diameter or surface coverage with simple water-derived adsorbates enhances the overall crystallinity of the NWs. NWs grown along the [010] direction are more ordered than those in the [001] direction. Both bare and hydrated NWs exhibit a band gap blue shift due to size confinement, but the shift varies with morphology and crystallinity. Hydrated NWs show a higher shift compared to their bare counterparts. The confined size benefits the electrochemical potential variations of TiO_2 NWs, favoring processes like water splitting. Surface coverage can also tune internal charge transport channels. The charge densities in the "bulk" band gap region involve surface atoms, indicating high photocatalytic activity. Surface hydration can optimize internal channels for photoelectron transport [21]. In addition, DFT has been used to investigate WO_3 nanowires [22], ZnO nanowires [23], Ge atomic nanowires [24], Si nanowires [25], and Mg nanowires [26].

8.3 Impact of Size and Shape on Electronic Properties

The size and shape of nanomaterials have significant effects on their electronic properties due to quantum confinement effects, surface-to-volume ratio, and shape anisotropy. As the size of the nanomaterial decreases to the nanoscale (typically below 100 nm), quantum confinement effects become significant. This leads to discrete energy levels rather than continuous energy bands seen in bulk materials. Smaller nanoparticles have larger band gaps because the energy difference between discrete levels increases as the size decreases. Quantum confinement can also be influenced by shape. For example, quantum dots (0D), nanowires (1D), and thin films (2D) exhibit different confinement in one or more dimensions, affecting their electronic properties. Quantum dots have confinement in all three dimensions, while nanowires have confinement in two dimensions, leading to different electronic behaviors. Smaller nanomaterials have a higher surface-to-volume ratio, which means a greater proportion of atoms are on the surface. This can significantly alter electronic properties due to surface states, defects, and interactions with the environment. Surface atoms often have different electronic properties compared to those in the bulk, affecting conductivity, reactivity, and other electronic behaviors.

The shape of nanomaterials can dictate the proportion of surface atoms and their arrangement. For instance, nanorods and nanoplates have different surface atom distributions compared to spherical nanoparticles, influencing their electronic properties differently. Anisotropic shapes, such as nanorods, nanowires, and nanoplates, exhibit directional dependence in their electronic properties. For example, nanowires can have higher conductivity along their length due to the preferential orientation of atoms and bonds. This anisotropy can be exploited in electronic devices where direction-dependent conductivity is desired. Both size and shape affect the density and type of surface states and defects, which can trap charge carriers or introduce localized states within the band gap. These surface states and defects play a crucial role in determining electronic properties such as charge carrier mobility, recombination rates, and conductivity. The electronic properties of nanomaterials also influence their optical properties. Quantum dots, for instance, exhibit size-dependent photoluminescence due to quantum confinement. The shape of nanomaterials can also lead to plasmonic effects, where conduction electrons oscillate in response to light, affecting absorption and scattering. The recombination and separation of charge carriers (electrons and holes) are influenced by the size and shape of nanomaterials. Smaller sizes can lead to faster recombination due to higher surface recombination rates, while certain shapes can promote efficient charge separation, beneficial for applications like photovoltaics and photocatalysis.

DFT simulations have been performed using the Amsterdam Density Functional (ADF) program. The gradient-corrected Becke–Perdew (BP86) exchange–correlation functional has been employed to optimize the molecular structures of all neutral and charged clusters. The article investigates the optical absorption properties of gold nanoparticles (Aun, n = 6–85) using TDDFT. Three structural shapes are considered: octahedra, truncated octahedra, and icosahedra, in various charge states corresponding to electronic shell closings. The BP86/DZ.4f level of theory has been used for structural optimization. The LB94 functional was employed for excitation energy calculations. Scalar relativistic effects have been incorporated using the zeroth-order regular approximation (ZORA). Excited states are calculated using TDDFT, with results convoluted with a Lorentzian function to produce smoothed spectra. Larger clusters show spectra less sensitive to charge state, whereas smaller clusters exhibit significant changes due to orbital occupation differences. Larger clusters' spectra are less affected by charge state changes, similar to the trend observed in octahedral clusters. Smaller clusters have sharper spectra and larger HOMO–LUMO gaps. These findings are consistent with other experimental and theoretical studies [27].

TDDFT has been applied to study gold clusters. Electronic structure has been investigated for different cluster sizes. The results could be explained using conventional quantum chemistry, analyzing excited states via one-electron configurations and molecular orbitals. However, for larger clusters, such analysis is impractical due to higher state densities and the overlap of s, p, and d bands near the Fermi level. Available experimental data on ligand-protected gold clusters align with these calculations. Recent advancements suggest using simplified schemes to solve TDDFT equations or performing dynamic polarizability calculations at complex frequencies based on the density, rather than the density matrix. This growing interest in nano-sized systems challenges quantum chemists to refine traditional algorithms and develop new methods to extend the applicability of accurate techniques [28].

DFT has been used to study the doping of silicon quantum dots (Si QDs) with boron (B) or phosphorus (P) to explore their potential in various applications such as tandem solar cells, thin-film transistors, and optoelectronic devices. Using DFT calculations with the CASTEP software, the researchers examined the structural and electronic properties of hydrogen-terminated Si QDs doped with a single impurity atom. Spin split occurs in some doped configurations, more pronounced in smaller QDs. Impurity positions affect impurity levels differently for B- and P-doping. B-doping causes deeper impurity states when moving from subsurface to center in smaller QDs, while P-doping causes shallower states. Spin splitting occurs in some configurations, with variations depending on QD size and impurity position. Smaller QDs exhibit more pronounced spin splits. The DOS analysis shows impurity-related peaks that shift depending on impurity position and QD size. B-doping peaks move toward the midgap region, while P-doping peaks move toward the conduction band as impurities shift from the subsurface to the center of the QDs [29].

References

1. T.A. Yousef, Structural, optical, morphology characterization and DFT studies of nano sized Cu(II) complexes containing schiff base using green synthesis. J. Mol. Struct. **1215**, 128280 (2020)

2. Li, Z., Ciobanu, C.V., Hu, J., Palomares-Ba´ez, J.P., Rodrı´guez-Lo´pez, J.L., Richards, R.: Experimental and DFT studies of gold nanoparticles supported on MgO(111) nano-sheets and their catalytic activity. Phys. Chem. Chem. Phys. 13:2582–2589 (2011)
3. H.F. El-Shafiy, M. Saif, M.M. Mashaly, S.A. Halim, M.F. Eid, A.I. Nabeel, R. Fouad, New Nano-complexes of Zn(II), Cu(II), Ni(II) and Co(II) ions; Spectroscopy, Thermal, Structural Analysis, DFT Calculations and Antimicrobial Activity Application. J. Mol. Struct. **1147**, 452–461 (2017)
4. Abdelghany, A.M., Sh. Mekhail, M., Abdelrazek, E.M., Aboud, M.M.: Combined DFT/FTIR structural studies of monodispersed PVP/Gold and silver nano particles. J. Alloy. Compd. **646**: 326–332 (2015)
5. Z. Parsaee, Sonochemical synthesis and DFT studies of nano novel schiff base cadmium complexes: green, efficient, recyclable catalysts and precursors of Cd NPs. J. Mol. Struct. **1146**, 644–659 (2017)
6. K. SenthilKannan, V. Sivaramakrishnan, V. Kalaipoonguzhali, M. Chinnadurai, S. Kannan, Electronic transport, HOMO–LUMO and computational studies of CuS monowire for nano device fabrication by DFT approach. Material Today: Proceedings **33**(7), 2746–2749 (2020)
7. N. Zare, A. Zabardasti, A. Mohammadi, F. Azarbani, Synthesis of spherical Fe_3O_4 nanoparticles from the thermal decomposition of iron (III) nano-structure complex: DFT studies and evaluation of the biological activity. Bioorg. Chem. **80**, 334–346 (2018)
8. H. Vovusha, B. Sanyal, DFT and TD-DFT studies on the electronic and optical properties of explosive molecules adsorbed on boron nitride and graphene nano flakes. RSC Adv. **5**, 4599–4608 (2015)
9. S. Banerjee, D. Bhattacharyya, Electronic properties of nano-graphene sheets calculated using quantum chemical DFT. Comput. Mater. Sci. **44**(1), 41–45 (2008)
10. M. Montazerozohori, S.A. Musavi, A. Masoudiasl, A. Hojjati, A. Assoud, DFT study and crystal structure analysis of a new nano-structure five coordinated Hg(II) complex involving C-H…O, N…O and $\pi \ldots \pi$ interactions in a supra-molecular structure. Spectrochim. Acta Part A Mol. Biomol. Spectrosc. **147**, 139–150 (2015)
11. Z. Tohidiyan, I. Sheikhshoaie, M. Khaleghi, J.T. Mague, A novel copper (II) complex containing a tetradentate Schiff base: synthesis, spectroscopy, crystal structure, DFT study, biological activity and preparation of its nano-sized metal oxide. J. Mol. Struct. **1134**, 706–714 (2017)
12. M.T. Baei, A.A. Peyghan, Z. Bagheri, A DFT Study on CO_2 Interaction with a BN Nano-Cage. Bull. Korean Chem. Soc. **33**(10), 3338–3342 (2012)
13. A. Raja, T. Adhikary, I.A. Al-Omari, G.P. Das, S. Ghosh, D.K. Satapathy, A. Oraon, J.E. Shield, S. Aich, Rapidly solidified Sm-Co-Hf-B magnetic Nano-composites: experimental and DFT studies. J. Magn. Magn. Mater. **504**, 166645 (2020)
14. M. Zhao, F. Yang, Y. Xue, D. Xiao, Y. Guo, A time-dependent DFT study of the absorption and fluorescence properties of graphene quantum dots. Chem. Phys. Chem **15**, 950–957 (2014)
15. X. Ji, A. Seif, J. Duan, A. Rashidi, Z. Zhou, S. Pourhashem, M. Mirzaee, X. Zhai, X. Zhao, B. Hou, Experimental and DFT studies on corrosion protection performance of epoxy/graphene quantum dots@TiO2 nanotubes coatings. Constr. Build. Mater. **322**, 126501 (2022)
16. O. Lehtonen, V.T. Sundholm, Computational studies of semiconductor quantum dots. Phys. Chem. Chem. Phys. **10**, 4535–4550 (2008)
17. C. Chou, C. Tseng, K.C. Pillai, B. Hwang, L. Chen, Surface related emission in CdS quantum dots. DFT simulation studies. J. Phys. Chem **115**, 20856–20863 (2011)
18. A.A.A. de Queiroz, M. Martins, D.A.W. Soares, E.J. França, Modeling of ZnS quantum dots synthesis by DFT techniques. J. Mol. Struct. **873**, 121–129 (2008)
19. S.K. Haram, A. Kshirsagar, Y.D. Gujarathi, P.P. Ingole, O.A. Nene, G.B. Markad, S.P. Nanavati, Quantum confinement in CdTe quantum dots: Investigation through cyclic voltammetry supported by density functional theory (DFT). J. Phys. Chem. C **115**, 6243–6249 (2011)
20. J. Singh, R. Thareja, P. Malik, R. Kakkar, Size-dependent structural and electronic properties of stoichiometric II-VI quantum dots and gas sensing ability of CdSe quantum dots: a DFT study. J. Nanopart. Res. **24**, 33 (2022)

21. A. Iacomino, G. Cantele, F. Trani, D. Ninno, DFT study on anatase TiO2 nanowires: structure and electronic properties as functions of size, surface termination, and morphology. J. Phys. Chem. **114**, 12389–12400 (2010)
22. Y. Qin, M. Liu, Z. Ye, A DFT study on WO_3 nanowires with different orientations for NO_2 sensing application. J. Mol. Struct. **1076**, 546–553 (2014)
23. M. Srivastava, A. Srivastava, Electron transport in CO_2 adsorbed ZnO nanowire: DFT study. Chem. Phys. Lett. **729**, 17–23 (2019)
24. D.E.P. Vanpoucke, G. Brocks, Formation of Pt-induced Ge atomic nanowires on Pt/Ge(001): A density functional theory study. Phys. Rev. B **77**, 241308 (2008)
25. Z. Guo, L.J. Arachchige, S. Qiu, X. Zhang, Y. Xu, S.J. Langford, C. Sun, P-Block element-doped silicon nanowires for nitrogen reduction reaction: a DFT study. Nanoscale **35**, 14935–14944 (2021)
26. L. Li, B. Peng, W. Ji, J. Chen, Studies on the hydrogen storage of magnesium nanowires by density functional theory. J. Phys. Chem. C **113**, 3007–3013 (2009)
27. G.T. Bae, C.M. Aikens, Time-Dependent density functional theory studies of optical properties of au nanoparticles: octahedra, truncated octahedra, and icosahedra. J. Phys. Chem. C **119**, 23127–23137 (2015)
28. M. Stener, A. Nardelli, R. De Francesco, G. Fronzoni, Optical excitations of gold nanoparticles: a quantum chemical scalar relativistic time dependent density functional study. J. Phys. Chem. **111**, 11862–11871 (2007)
29. J. Zeng, H. Yu, A first-principle study of B- and P-doped silicon quantum dots. J. Nanomater. (2012). https://doi.org/10.1155/2012/147169

Chapter 9
DFT in Defects and Disorder

Hari Krishna Neupane and Narayan Prasad Adhikari

Abstract Crystal defects play a crucial role in determining the physical, chemical, and electronic properties of materials. This study presents a comprehensive overview of different types of defects in solids, including point defects, line defects, and surface defects, along with their classification and fundamental characteristics. This work highlights the role of density functional theory (DFT) in studying vacancy defects at the atomic level. It also covers line defects such as dislocations and surface defects that strongly influence material properties. The application of DFT in disordered systems, including modeling techniques and practical uses, is discussed. Additionally, impurity states and doping effects are examined using DFT-based approaches. In essence, the study emphasizes defect engineering and computational methods for understanding and tailoring advanced material properties.

9.1 Introduction

Regular arrangement of atoms is an important property of crystal. Based on the arrangement of atoms in the crystal, crystals are generally classified into two types. They are ideal crystal and real crystal. Ideal crystals are formed by the sum of lattice and motif. The three-dimensional (3D) periodic arrangement of atoms or ions in the crystal is called lattice, while an atom or group of atoms in crystal is called motif. In an ideal crystal, the atoms are assumed to be arranged in a perfectly regular manner. Such a concept of an ideal crystal is used to explain various intensive properties of the lattice structure. For example, specific heat, density, elasticity, magnetic properties, etc. [1, 6].

The crystals which are deviated from their ideality due to the presence of defects or imperfections are called real crystals. All the actual crystalline materials are not

H. K. Neupane · N. P. Adhikari (✉)
Central Department of Physics, Institute of Science and Technology, Tribhuvan University, Kirtipur, Kathmandu, Nepal
e-mail: narayan.adhikari@cdp.tu.edu.np

H. K. Neupane
e-mail: hari.neupane@ac.tu.edu.np

R. Thotagamuge (ed.), *Density Functional Theory*,
https://doi.org/10.1007/978-981-95-9745-1_9

perfect. A defect in a solid is any deviation from the periodicity in the solid. The physical and mechanical properties of crystals are controlled by its defects. The defects influence the properties of the solids in the following ways:

i. The defects may scatter conduction electrons in a metal, increasing an electrical resistance of the metal.
ii. Atomic diffusions may be accelerated enormously by the impurities are imperfections.
iii. Some defects even if present sparingly decreases the strength of crystals.
iv. Pure crystals having impurities and imperfection are offer colored.

The common point defects (imperfections) in crystals are chemical impurities, vacant lattice sites, and extra atoms not in regular lattice positions. All crystals must have some type of lattice defects. For many purposes, the defect may be neglected but at other times, lattice defects are considered as important position in Modern Solid Physics [1, 6].

In thermal equilibrium, certain number of lattice vacancies are always presented in the crystal; this is because the periodic arrangement of the atoms in crystal is changed due to the effect of temperature, and defect arises. The real crystal has temperature more than absolute temperature, so it has lattice defect. Real crystals are not existed at absolute temperature (T = 0 K), but they have any finite temperature, i.e., above T = 0 K, defect arises. The crystal is existed at absolute temperature (0 K) refers the ideal crystal, at which atoms are arranged in periodic arrangement, and can't exist defect. Hence, real crystal is always imperfect.

9.2 Classification of Defects

Based on the dimensionality, defects in crystals are classified into three types. They are:

i. Zero-dimensional or point defects.
ii. One-dimensional or line defects.
iii. Two-dimensional or surface defects. They are described as following.

9.3 Point Defects

Point defects are single-point disruptions in a crystalline materials, in which arrangement of atoms or ions take place at a single location inside the lattice. These defects arise from structural or asymmetrical imperfections [6, 9]. They have a significant-effect on the material's properties such as strength, conductivity, and optical properties. Point defects are generally three types. They are substitutional defects, vacancy defects, and interstitial defects.

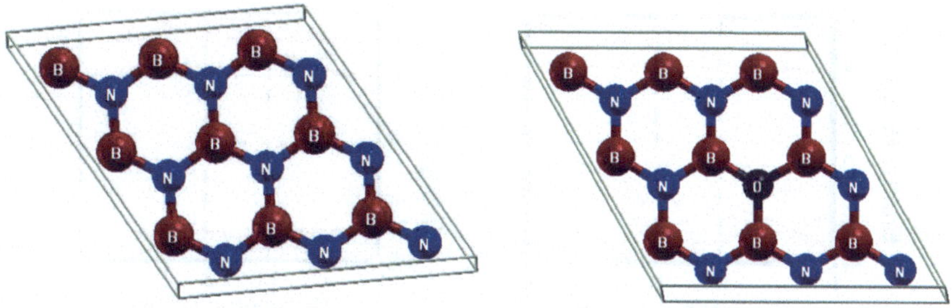

Fig. 9.1 **a** Crystal structure of hexagonal Boron Nitride (h-BN), **b** crystal structure of substitutional defected h-BN, where one Nitrogen atom (N) is replaced by one Oxygen (O) atom in the crystal structure

9.3.1 Substitutional Defects

An atom or ion in the lattice is replaced by a different type of atom or ion is called substitutional defects [6]. It means, a new kind of atom or ion takes place of an existing one in the lattice. Substitutional defects are also known as impurity defects, in which regular arrangement of atom in the crystal structure is replaced by foreign atom. In the lattice structure of hexagonal Boron Nitride (h-BN) is shown in Fig. 9.1a, where Boron (B) and Nitrogen (N) atoms are covalently bounded in the lattice sites. The one N atom is replaced by one Oxygen(O) atom form substitutional defects in the structure as shown in Fig. 9.1b.

From the Density Functional Theory (DFT) methods of calculations, it is found that electronic properties of h-BN are predicted based on the band structure calculations and found that it is a wide band gap material of band gap energy 4.98 eV [12], which is shown in Fig. 9.2a. A substitutional defected crystal structure of h-BN is shown in Fig. 9.1b, which is formed by the replacement of one N atom by one O atom in structure. It has 4.19 eV band gap energy [12], which is shown in Fig. 9.2b. It means, the band gap energy of pristine material is changed due to the presence of substitutional (impurity) defect in structure. Hence, impurity defect can affect the material's properties.

9.3.2 Vacancy Defects

The simplest imperfection is a lattice vacancy, which is a missing atom or ion is known as vacancy defect in a lattice crystal. In the crystal, there may be some vacant sites. The vacancy defect must be at there. It means, there must be a certain fraction of vacant sites in a crystal at thermal equilibrium. The thermal equilibrium means, minimum Gibbs free energy (G) at constant temperature (T), and pressure (P). A crystal with vacancies has a lower free energy G than a perfect crystal. There will be

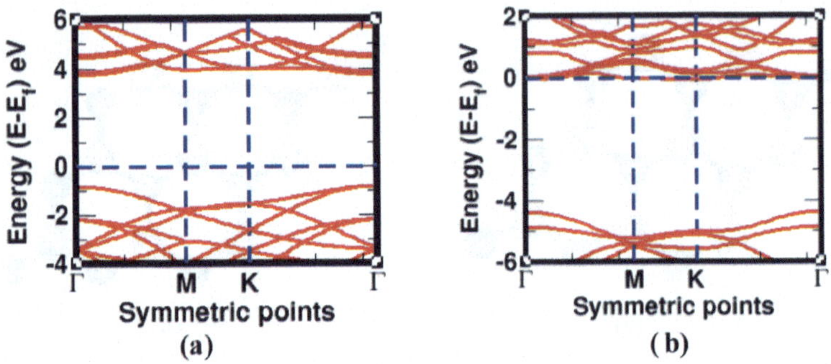

Fig. 9.2 **a** Band structure of hexagonal Boron Nitride (h-BN), **b** band structure of substitutional defected h-BN where, one Nitrogen atom (N) is replaced by one Oxygen (O) atom in the crystal structure, and the horizontal dot line represents the fermi energy level which separates the valence band and conduction band. The region above the fermi energy level is called conduction band and below the fermi energy level is called conduction band [12]

certain equilibrium concentration of vacancies which will minimize the free energy in the crystal [6]. One of the perfect examples of vacancy defect is Schottky defect and Frenkel defects, which is explained as following.

Schottky Defect

A Schottky defect is created in a perfect crystal when an atom or ion from a lattice site in the interior transferred to a lattice site on the surface of the crystal [6]. In thermal equilibrium, a certain number of lattice vacancies are always present in an otherwise perfect crystal because the entropy is increased by the presence of disorder in the structure. The Schottky defect in a monatomic crystal is shown in Fig. 9.3.

Let n and N are the number of vacancies and total number of atoms in a crystal, respectively. E_v be the energy required to create a vacancy then, total possible ways to take N atoms for n lattice points is

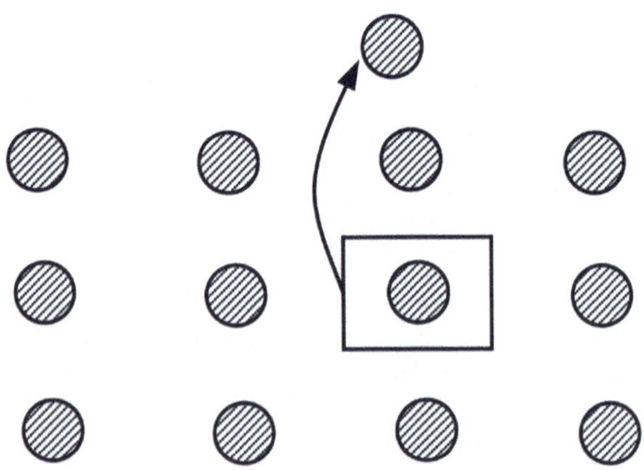

Fig. 9.3 Schottky defect in a crystal, where one atom from the lattice sites is shifted on its surface

$$\Omega = \frac{N!}{N!(N-n)!}. \tag{9.1}$$

Then, the disorder increases due to the creation of "n" vacancies. The corresponding increase in entropy, which is given by

$$S = K_B ln\Omega,$$

$$S = K_B \ln|\frac{N!}{n!(N-n)!}|. \tag{9.2}$$

This ultimately changes Gibbs free energy (G), which is given by

$$G = H - TS, \tag{9.3}$$

where $H = E + PV =$ enthalpy of energy, and E, P, V, and S are internal energy, pressure, volume, and entropy, respectively. Using Eq. (9.2) in Eq. (9.3), we get

$$G = nE_V - K_B T ln\left|\frac{N!}{n!(N-n)!}\right|,$$

$$G = nE_V - K_B[lnN! - lnn! - \ln(N-n)!].$$

Using Stirling approximation, $lnx! = xlnx - x$, we get

$$G = nE_V - K_B T[NlnN - N - nlnn + n - (N-n)\ln(N-n) + N - n],$$

$$G = nE_V - K_B T[NlnN - nlnn - (N-n)\ln(N-n)].$$

In thermal equilibrium, the Gibbs free energy must be minimum at temperature T with respect to changes in "n", then,

i.e.,

$$\frac{dG}{dv}|_T = 0 = E_V - K_B T\left[-lnn - \frac{1}{n}.n + \ln(N-n) + (N-n).\frac{1}{(N-n)}\right]$$

or,

$$0 = E_V - K_B T[-lnn - 1 + \ln(N-n) + 1]$$

or,

$$E_V - K_B T\left[\ln\left(\frac{N-n}{n}\right)\right] = 0$$

or,

$$\ln\left(\frac{n}{N-n}\right) = -\frac{E_V}{K_B T}$$

or,

$$n = (N-n)e^{-\frac{E_V}{K_B T}} \tag{9.4}$$

since, N > > n,

$$n = Ne^{-\frac{E_V}{K_B T}}. \tag{9.5}$$

This gives the equilibrium concentration of vacancies, due to Schottky defect at temperature T. This relation shows that the equilibrium concentration of vacancies that is number of defects increases exponentially as its temperature increases. Furthermore, a certain number of defects are always present at all temperature above the absolute zero. Here, a real crystal is always imperfect. The free energy is smaller than zero, so it is thermodynamic necessity. Figure 9.4 also shows that real crystal is always imperfect.

Frenkel Defect

Frenkel defect is a type of vacancy defect, in which an atom is transferred from a lattice site to an interstitial position, and then a vacancy is created. This type of defect in crystal is known as "Frenkel detect" [6]. The interstitial atoms are extra atoms squeezed into the spaces between normal atom sites, as shown in Fig. 9.5.

We can calculate the number of Frenkel defects in equilibrium at any temperature T. Let E_i be the energy required to displace atoms or ions from a regular lattice site of a crystal to an interstitial position. N' be the interstitial atoms and N be the total number

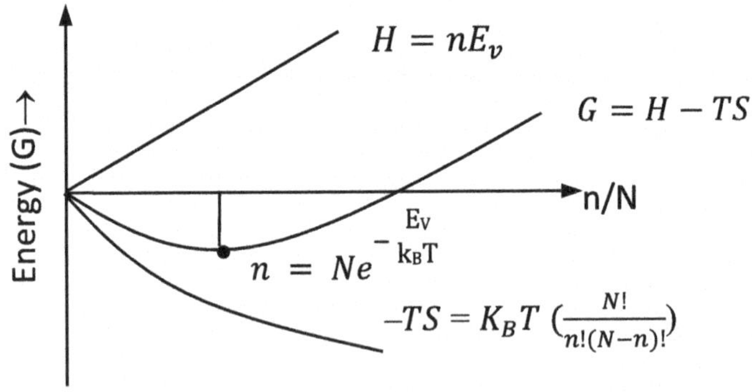

Fig. 9.4 Variation of Gibbs free energy with n/N

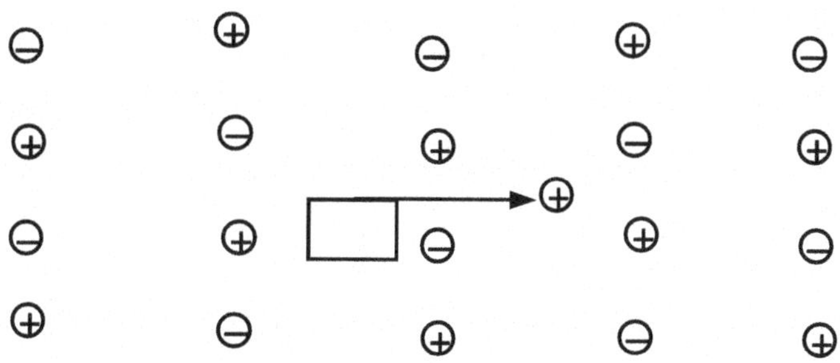

Fig. 9.5 Frenkel defect in an ionic crystal

of atoms, then the number of ways in which “n” Frenkel defects can be formed is given by

$$\Omega = \frac{N!}{N!(N-n)!}\frac{N'!}{N!(N'-n)!}.$$

Entropy is increased due to disorder of atoms or ions in the crystal structure. Therefore,

$$S = K_B ln\Omega = K_B \ln\left[\frac{N!}{N!(N-n)!}\frac{N'!}{N!(N'-n)!}\right]$$

which in turn produces a change in Gibbs free energy

$$G = H - TS,$$

$G = nE_i - K_B\ln\left[\frac{N!N'!}{(n!)^2(N-n)!(N'-n)!}\right]$,

$G = nE_i - K_BT\ [lnN! - ln(N-n)! - 2lnn! + lnN'! - ln(N'-n)!.$

Using Stirling approximation,

$G = nE_i\ - K_BT\ [NlnN - N - (N-n)ln(N-n) + N - n - 2nlnn + 2n + N'lnN'$ $- N' - (N'-n)ln(N'-n) + N' - n].$

$G = nE_i - K_BT\ [NlnN - (N-n)ln(N-n) - 2nlnn + N'lnN' - (N'-n)ln(N'-n)].$

At equilibrium, Gibbs free energy will be minimum. Thus,

$$\frac{\partial G}{\partial n}|_T = 0 = E_i - K_BT[ln(N-n) - (N-n).\frac{(-1)}{(N-n)}$$
$$- 2lnn - 2 + ln(N'-n) - \frac{(N'-n)(-1)}{(N'-n)}]$$

$$0\ =\ Ei\ - K_BT\left[ln(N-n) + 1 - 2lnn - 2 + ln(N'-n) + 1\right]$$

$$\ln[\frac{(N-n)(N'-n)}{n^2}] = \frac{E_i}{K_B T}$$

$$\ln[\frac{n^2}{(N-n)(N'-n)}] = -\frac{E_i}{K_B T}.$$

If $N' >> n$ and $N >> n$, then,

$$n2 = (N-n)(N'-n)e$$

$$n = \sqrt{\mathrm{NN'}} \cdot \mathrm{e}^{-\frac{\mathrm{E_i}}{2\mathrm{k_B T}}}$$

It is obvious from the relation that the Frenkel defects at equilibrium increase with increase in temperature. Above absolute zero, this also proves that a real crystal is always imperfect.

9.3.3 DFT Approach on Vacancy Defects

In Condensed Matter Physics and Material Science research, Density Functional Theory (DFT) method was used to predict the materials physical properties. Vacancy defects tune the noble properties in the materials [9, 12]. In computational Physics, DFT method was widely used for the prediction of material's properties. We have taken an example of computational research works on vacancy defected materials by employing DFT methods, which is given as following.

Example 1 The bilayers supercell structure of hexagonal Boron Nitride (h-BN) is constructed, which is shown in Fig. 9.6a. In that structure, Boron and Nitrogen atoms are covalently bounded to each other. The weak (van der Waal) force is existed in between the interlayer of bilayer structure. The Nitrogen (N) atom vacancy defected bilayer supercell h-BN structure is formed by removing one N atom from the upper layer of BN site is shown in Fig. 6b. The binding energy and total ground state energy of defected material has altered due to the effect of vacancy defected atom, which changes the material's physical (electronic and magnetic) properties.

The electronic properties of considered materials are predicted based on their band structure calculations, which are shown in Fig. 9.7a and b. In Fig. 9.7a, it is seen that band gap energy of bilayer h-BN is found to be 4.02 eV [8]. It reflects that it is wide band gap (insulating) material. One N atom vacancy defect is created in pristine h-BN, its band structure is slightly different than the band structure of pristine h-BN material. The band gap energy of one N defected h-BN (1N_h-BN) is found to be 1.80 eV [8], which is lower than the band gap energy of pristine h-BN material. It reveals that defected material has semiconducting properties. The insulating properties of material are changed into semiconducting properties of material

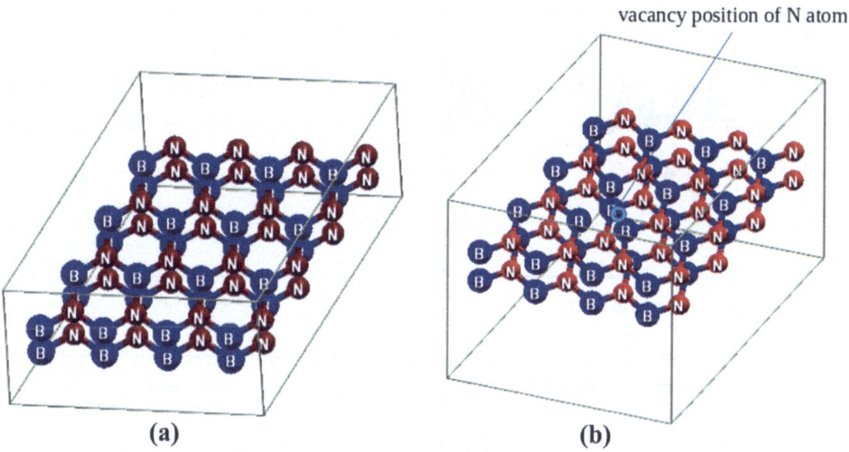

Fig. 9.6 **a** Crystal structure of bilayer h-BN supercell, **b** one N atom vacancy defected h-BN crystal structure

due to the presence of vacancy defect. Hence, from the DFT calculations, material's electronic properties found to be altered due to the presence of vacancy defects in the structure.

Moreover, the magnetic properties of materials are examined by the analysis of their Density of States (DOS) and Partial Density of States (PDOS) calculations. The DOS and PDOS plots of h-BN are shown in Fig. 9.8a and b and of 1N_h-BN are illustrated in Fig. 9.8c and d, respectively. It is seen that up-spin and down-spin states of electrons in the orbital of atoms are symmetrically distributed in both DOS and PDOS plots of h-BN. It means, it has no magnetic moment. Hence, h-BN is a non-magnetic material. On the other hands, up-spin and down-spin states are asymmetrically distributed around the fermi energy level in DOS and PDOS plots of 1N_h-BN material. It implies that one N atom vacancy defected h-BN has magnetic moment of value 1.00 Bohr magneton per cell [8]. Hence, 1N_h-BN has magnetic

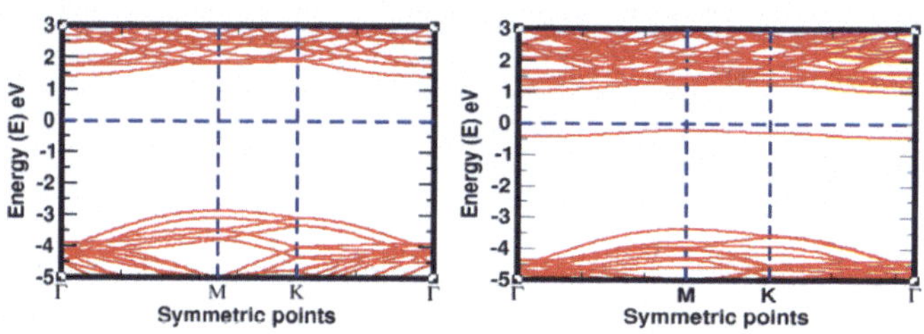

Fig. 9.7 **a** Band structure plot of bilayer supercell structure of h-BN, **b** band structure of one N defected bilayer supercell structure of h-BN. Where horizontal dot line represents the fermi energy level [8]

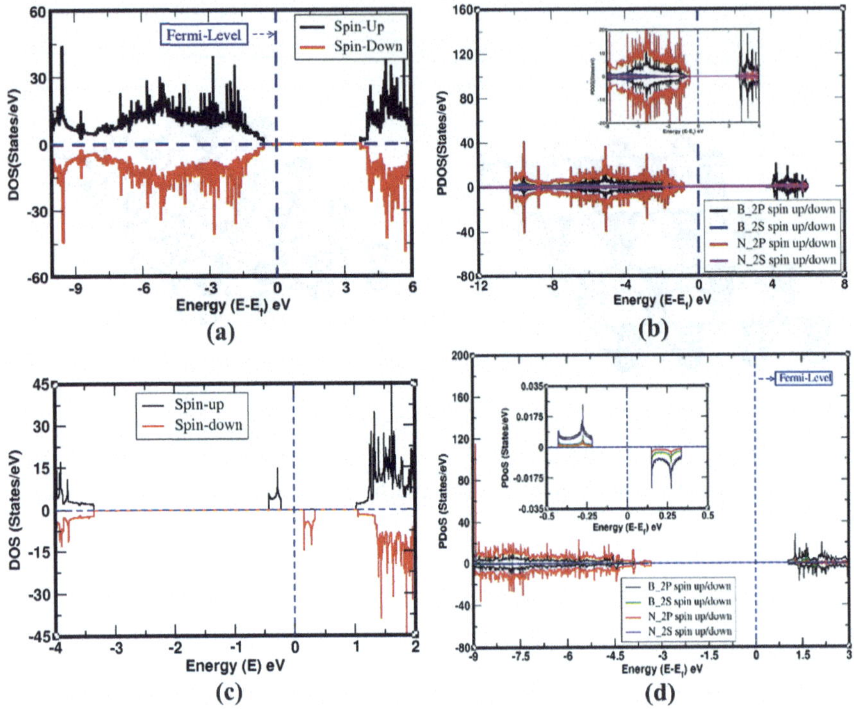

Fig. 9.8 **a** DOS plot of bilayer supercell structure of h-BN, **b** PDOS plot of bilayer supercell structure of h-BN, **c** DOS plot of bilayer supercell structure of 1N_h-BN, and **d** PDOS plot of bilayer supercell structure of 1N_h-BN. The vertical dot line represents the fermi energy level in all plots [8]

properties. Thus, non-magnetic material changes into magnetic material due to the presence of vacancy defect in the structure.

In conclusions, from the DFT method of calculations, it is found that material physical properties can be altered due to the presence of vacancy defects in the structure. Hence, vacancy defects tune the unique properties in the materials.

9.4 Line Defects

The defects that affect a row or line of atoms in the lattice are called line defects [6]. Line defects can be explained in term of dislocations which are explained as following.

9.4.1 Dislocations

The low observed values of the critical shear stress are explained in terms of motion through the lattice of a line imperfection known as a dislocation. The idea that slips propagates by the motion of dislocations were introduced independently by different scientists [4, 7, 16]. There are several basic types of dislocations,among them two of them have different importance. They are given as following.

i. **Edge Dislocation**

If the dislocation occurs at the lower edge of extra atomic plane, this type of dislo cation is called edge dislocation and is caused by inserting an extra plane of atoms on the upper half of the crystals. This type of edge dislocation is called positive edge dislocation. On the other hand, if extra plane of atoms is inserted below the slip plane, then this type is called negative edge dislocation [7]. A simple cubic crystal in which slip of one atom distance has occurred over the left half of the slip plane but not over the right half as shown in Fig. 9.9. An edge dislocation EF in the smooth plane ABCD in Fig. 9.9a shows the slipped region ABEF in which the atoms have been displaced by more than half a lattice constant and the un-slipped region FECD with displacement less than half a lattice constant. The boundary between the slipped and unslipped regions is called the dislocation. Its position is marked by the termination of an extra vertical half plane of atoms crowded into the upper half of the crystal as shown in Fig. 9.9b.

Near the dislocation, the crystal is highly strained. The simple edge dislocation extends indefinitely in the slip plane in a direction normal to the slip direction. The mechanism responsible for the mobility of a dislocation is shown in Fig. 9.10.

If atoms on one side of the slip plane are moved with respect to those on the other side, atoms at the slip plane will experience repulsive forces from some neighbors and attractive forces from others across the slip plane. These forces cancel to a first

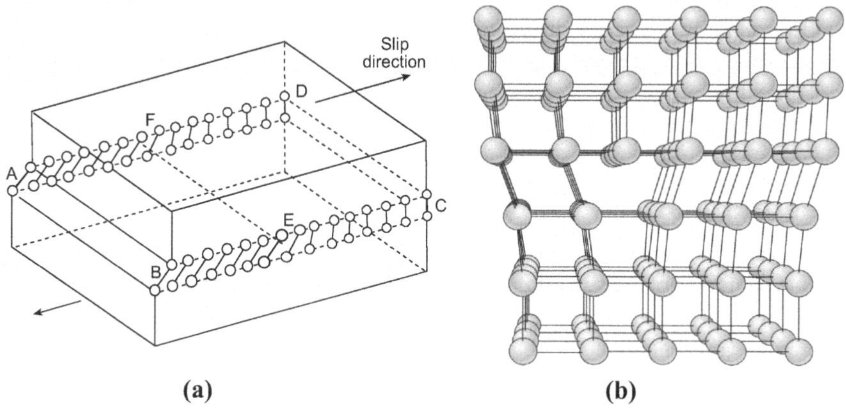

Fig. 9.9 **a** Edge dislocation, **b** structure of edge dislocation

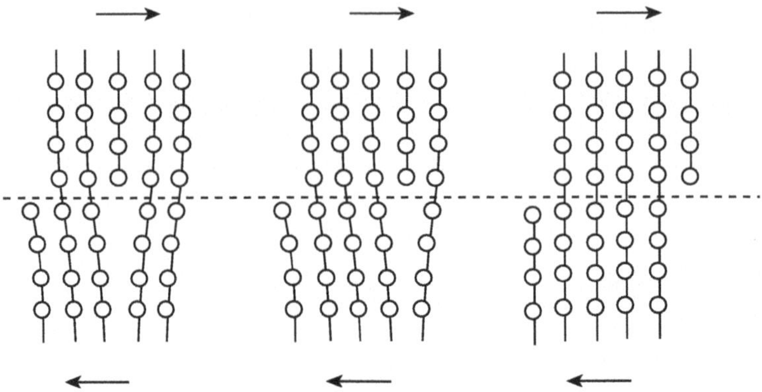

Fig. 9.10 Motion of dislocation under a shear tending to move the upper surface of the specimen to the right

approximation. The external stress required to move a dislocation has been calculated and is quite small, probably below 10^5dyne/cm^2, provided that the bonding forces in the crystal are not highly directional [4, 7, 16]. Thus, dislocations may make a crystal very plastic. Passage of a dislocation through a crystal is equivalent to a slip displacement of one part of the crystal.

ii. **Screw Dislocation**

The screw dislocation is shown in Fig. 11a and b, respectively. A screw dislocation marks the boundary between slipped and unslipped parts of the crystal. The boundary parallels the slip direction, instead of lying perpendicular to it as for the edge dislocation. The screw dislocation may be thought of as produced by cutting the crystal partway through with a knife and shearing it parallel to the edge of the cut by one atom spacing. A screw dislocation transforms successive atom planes into the surface of a helix; this accounts for the name of the dislocation [4, 7, 16]. In Fig. 9.11a, a part ABEF of the slip plane has slipped in the direction parallel to the dislocation line EF. A screw dislocation may be visualized as a helical arrangement of lattice planes, such that we can change the planes ongoing completely around the dislocation line.

9.5 Surface Defects

A surface defects are the two-dimensional interface between adjacent grains (single crystal) in a polycrystalline material. The surface defects are described in terms of free surface, grain boundary, twin boundary, and stacking fault [6].

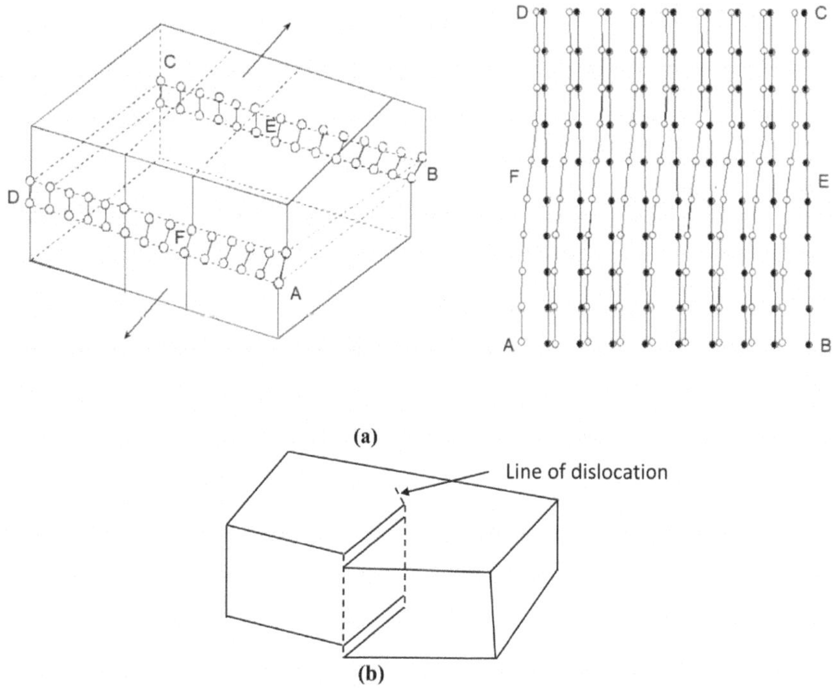

Fig. 9.11 **a** A screw dislocation, **b** screw dislocation, the broken vertical line that makes the dislocation is surrounded by strained material

9.6 DFT in Disordered Systems

The disordered systems are materials or structures that lack long-range order or periodicity.

DFT is a powerful tool for understanding disordered systems, though computational technique. Its application is constantly growing due to advancements in techniques, algorithms, and computational resources, which offer crucial insights into the links between structure and properties of disordered materials. DFT has some techniques to model disordered systems. They are discussed as below.

9.6.1 Techniques to Model Disordered Systems

To model the disordered systems, the following steps are required.

i. **Prepared the Supercell Structure**

The supercell structure is prepared by using DFT technique, which represent the disordered system.

ii. **Defect Modeling**

The defects are introduced in the supercell structure of crystalline material. The structure optimization and relax calculations are done of defected supercell structure.

iii. **Computational Approach**

To analysis the physical properties (for better accuracy in electronic structure calculations) of disordered materials, we can use the GGA with PBE or GGA with PBE + U or others hybrid functionals in DFT method of calculations [5]. For the calculations of material's physical properties, first choose the functional and then optimize the k-points, lattice constant, and kinetic energy cutoff parameters. These optimized values are used in the input files, then relax calculations are done by the convergence test. After relaxed the structure, further calculations can be done for the investigation of materials noble properties. In DFT methods, various software packages are used for the computational purposes, such as; VASP (Efficient for supercell calculations and disordered systems) [3], Quantum ESPRESSO (Open-source software with versatile functionality) [2], ABINIT (suitable for various types of materials, including disordered systems); WIN2K (for the crystalline or molecular system); CP2K (well-suited for large-scale and molecular system), etc. [17].

9.6.2 Uses of DFT in Disordered Systems

DFT method can be used to explore some important physical properties of defected materials. Some important properties of materials are given as following:

i. **Structural Properties**

The structural properties of materials are examined by the estimation of their ground states energy, binding energy, pair correlation functions, bond angle distributions, radial distribution functions, etc., by using DFT methods of calculations [10].

ii. **Electronic Properties**

The electronic properties of materials are predicted by the analysis of their band structures and Density of States (DOS) and localization of electronic states calculations [11].

iii. **Magnetic Properties**

The magnetic properties of materials are studied by the analysis of materials Density of States (DOS) and Partial Density of States (PDOS) plots [12].

iv. **Mechanical and Dynamical Properties**

The mechanical properties of materials are examined by the estimation of their elastic constants and modulus of rigidities. The dynamical stability of materials is studied by the analysis of materials phonon dispersion curves [15].

v. **Thermodynamical Properties**

Thermodynamical properties of materials are studied by calculations of material's phonon velocities, specific heat capacity, free energy, entropy, and Debye temperature [15].

vi. **Optical Properties**

The optical properties of materials are examined by development of real and imaginary dielectric function, refractive index, reflection coefficient, absorption coefficient, extinction coefficient, energy loss with photon energies, etc. [15].

vii. **Transport Properties**

Transport properties of materials are investigated by the calculations of their Seebeck coefficient, electrical conductivity, thermoelectric power factor, thermal conductivity, and figure of merit, and then find the thermal efficiency of materials [13].

9.6.3 Example of DFT Calculations on Disordered Systems

Here, we discussed the disordered system like one N atom vacancy defected Graphene and hexagonal Boron Nitride (G/h-BN) heterostructure (HS) material. This material is prepared by removing one N atom from the surface of BN site of HS material. Then, we have studied various physical properties of that material. Structure preparation and computational investigation of its various properties by using DFT methods are discussed as following.

Structure Optimization

To study the physical properties (like electronic, magnetic, optical, transport properties, etc.,) of a material, it needs to optimize the material's kinetic energy cutoff, lattice parameter, and k-points. They are discussed as following.

To find kinetic energy cutoff, we performed the self-consistent functional calculations (scf) by using experimental values of lattice parameter (a) and arbitrary value of k-points for different values of energy cutoff (like: 10, 15, 20, 25, 30, 35, 40, 45, 50.... etc.). Finally, we obtained ground state energy for each value of energy cutoff using executable 'pw.x' command. Then, graph is obtained between total energy and corresponding energy cutoff value, which is shown in Fig. 9.12a, where the optimized value of energy cutoff is found to be 35 Ry. In order to optimize k-points, we used energy cutoff value = 35 Ry, and experimental value of lattice parameter in the input file, and then performed scf calculations for different value of k-points (like: 6, 8, 10, 12, 14, 16, 18, 20, 22, 24, 26, 28, 30,... etc.). Then, we obtained total energy for each value of k-points. The graph is obtained by plotting the values of total energy with number of k-points, and found total energy value is constant at constant value of k-points (i.e., K = 22) as shown in Fig. 9.12b. That constant k-point value is called optimized k-point value, which is used for further calculations. After the optimization of energy cutoff and k-points, we proceed to find the optimized value of lattice parameter. For this, we used the optimized value of energy cutoff and k-points, then performed scf calculations for different value of lattice parameter (like: 4.40, 4.45, 4.50, 4.55, 4.60, 4.65, 4.70, 4.75, 4.80, 4.85, 4.90, 4.95,... etc.,). Finally, we got total energy for each value of lattice parameter, and then plot a graph of total energy versus lattice parameters, which is shown in Fig. 9.12c. The lattice parameter is noted 4.65 Bohr at minimum value of total energy. This is the optimized value of lattice parameter. These optimized values of energy cutoff, k-points, and lattice parameters are taken in the input scf file, and then relax calculations are done to get actual values of atomic coordinates of atoms in the structure.

These optimized values of energy cutoff, k-points, and lattice parameters are taken in the input scf file; then relax calculations are done to get actual values of atomic coordinates of atoms in the structure. That structure is called optimized and relax structure of material, which can be used for the investigation of material's properties. An optimized and relaxed structure of one N atom vacancy defected G/h-BN (1N_G/h-BN) is shown in Fig. 9.13.

Electronic and Magnetic Properties

The electronic properties of 1N_G/h-BN is examined based on its band structure and Density of States (DOS) plots, which are shown in Fig. 9.14a and b, respectively. From these plots, it is found that 1N_G/h-BN has metallic properties this is because electronic band states of atoms are presented around the fermi energy level. On the other hand, the Density of States (DOS) and Partial Density of States (PDOS) plots of 1N_G/h-BN, shown in Fig. 9.14b and c, respectively, are analyzed. It is found that material has magnetic properties. The magnetic properties in the material have developed due to the asymmetrical distribution of up-spin and down-spin states of electrons in the orbital of atoms present in the material [13].

In the conclusions, the different physical properties (like electronic, magnetic, optical, etc.) of disordered systems are investigated computationally by using DFT methods employing LDA with PBE, GGA with PBE, and others hybrid functional.

9.7 Impurity States and Doping

A material develops impurity states when its periodic crystal structure is disrupted by foreign atoms or structural defects. Impurity states can change the optical, magnetic, and electronic properties of materials [13]. The host atoms at lattice sites are replaced with foreign atoms is called substitutional impurity states, while the spaces between the lattice atoms are occupied by foreign atoms in the crystal is called interstitial impurity states. A Silicon atom is replaced by a Phosphorus atom in the lattice sites of crystal is an example of substitutional impurity states.

An intentionally introduce the impurity atoms in the material in order to alter a material's electronic properties is called doping. It is a crucial technique in semiconductors that allows for the regulation of carrier type and concentration. Donor

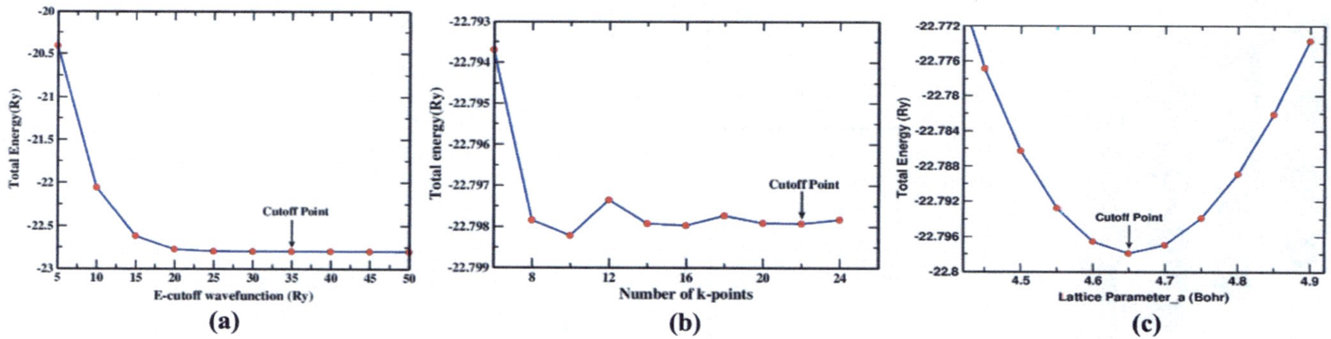

Fig. 9.12 **a** Graph between total energy cutoff versus total energy, **b** graph between number of k-points versus total energy, and **c** graph between lattice parameter vs. total energy

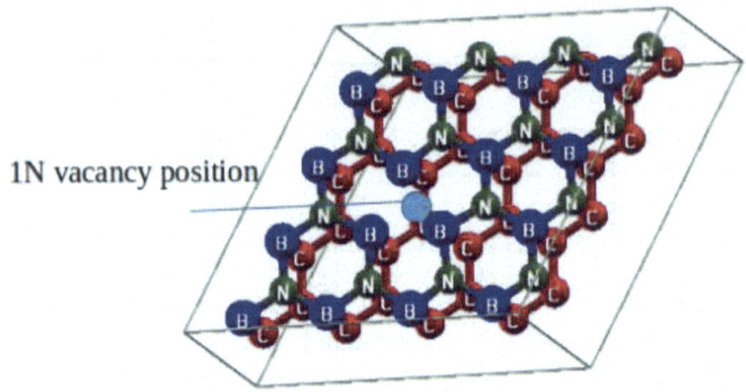

Fig. 9.13 Optimized and relaxed structure of one N atom vacancy defect on G/h-BN heterostructure material

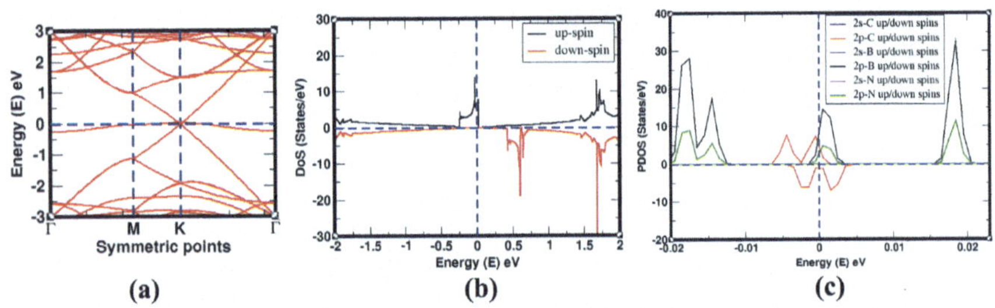

Fig. 9.14 **a** Band structure of 1N_G/h-BN, **b** DOS plot of 1N_G/h-BN, and **c** PDOS plot of 1N_G/h-BN [13]

impurities with extra electrons are doped in the crystal structure which increases the electron carriers is called n-type of doping, e.g., Phosphorus is doped in Silicon, while acceptor impurities atoms doped in the crystal structure is called p-type doping, e.g., Boron is doped in Silicon. Impurity states and doping can change the materials physical and chemical properties.

9.7.1 DFT in Impurity States and Doping

In this section, we discussed the physical properties of impurity defected system and foreign atom or molecule doped on the impurity defected system by using DFT methods of calculations. Firstly, we discussed the DFT methods of calculations in impurity states of material. Secondly, we discussed the uses of DFT methods in doping system.

9.7.2 DFT Methods in Impurity States

One Tc impurity defected MoS_2 material is prepared by replacing the one Mo atom by one Tc atom in the supercell structure of Molybdenum-disulfide (MoS_2) as shown in Fig. 9.15.

At first, optimized and relaxed supercell structure of MoS_2 is prepared through DFT methods of calculations by employing quantum ESPRESSO software. Then, one Tc impurity atom defected MoS_2 material is prepared by replacing; one Mo atom by one Tc atom in the structure, which is shown in Fig. 9.15. That defected structure is again relaxed and found to be stable structure. The stability of that structure is examined by finding its ground state energy and binding energy. It is found that one Tc impurity defected MoS_2 is a stable material structure [14]. The electronic properties of considered material are investigated by the interpretation of it band structure and Density of States (DOS) plots, which are shown in Fig. 9.16a and b, respectively. The magnetic properties of the material are predicted by using the DOS and Partial Density of States (PDOS) plots; they are illustrated in Fig. 9.16b and c, respectively.

From the analysis of band and DOS plots, defected material has semiconducting properties. Its band gap energy is found to be nearly 1.20 eV [13]. From the analysis

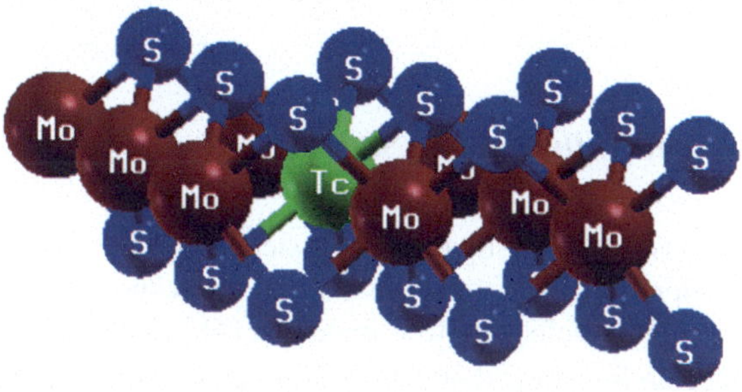

Fig. 9.15 One Tc atom impurity defected MoS_2 supercell structure

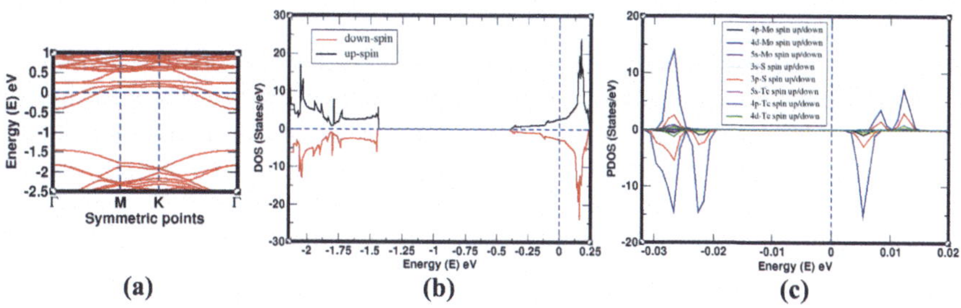

Fig. 9.16 **a** Band structure of Tc defected MoS_2, **b** DOS plot of Tc defected MoS_2, and **c** PDOS plot of Tc defected MoS_2 material. In band structure plot, horizontal dot line indicates the fermi energy level while in DOS and PDOS plots, vertical dot line indicates the fermi energy level [13]

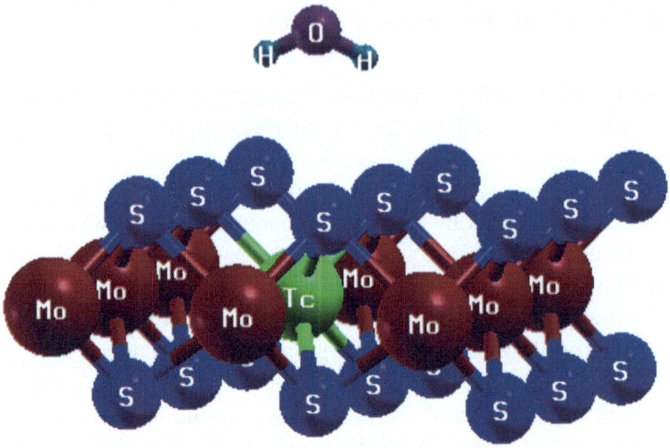

Fig. 9.17 Water doped on one Tc atom impurity defected MoS_2 supercell structure

of DOS and PDOS plots, defected material has magnetic properties, which is due to the asymmetrical distributed up-spin and down-spin states of electrons in the orbital of atom present in the material [14]. The mentioned above properties of material are examined on the basis of DFT methods of calculations.

9.7.3 DFT Methods in Doping States

DFT methods also used to investigate the physical properties of doping states of materials computationally. Here, we discussed the water molecule is doped on the surface of one Tc impurity defected MoS_2 material [14], which is shown in Fig. 9.17.

The water molecule is adsorbed at 2.45 Å distance on the surface of Tc defected MoS_2 material [14]. From the structural analysis, it is found to be stable material. The electronic and magnetic properties of water doped Tc defected MoS_2 material through its band-DOS plots and DOS-PDOS plots, respectively, which are illustrated in Fig. 9.18a, b, and c.

From the band and DOS plots, it is found that water doped on Tc defected MoS_2 has semiconducting properties of band gap energy is found to be 1.02 eV (Neupane et al., 2023), which is slightly less than the band gap energy of only Tc defected MoS_2. It reflects that doping molecule affects the properties of material. On the other hand, magnetic properties of considered material are estimated by using DFT methods of calculations, it is found to be magnetic material. Hence, multi-properties of defected materials can be explored through DFT methods of calculations computationally.

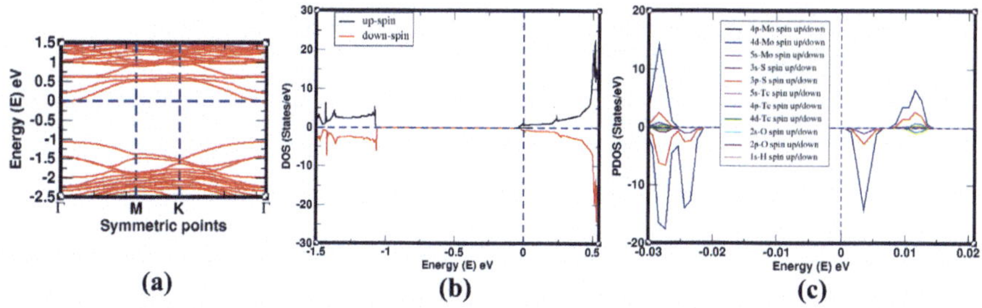

Fig. 9.18 **a** Band structure of water doped on Tc defected MoS_2, **b** DOS plot of water doped on Tc defected MoS_2, and **c** PDOS plot of water doped on Tc defected MoS_2 material. In band structure plot, horizontal dot line indicates the fermi energy level, while in DOS and PDOS plots, vertical dot line indicates the fermi energy level [14]

References

1. N.W. Ashcroft, N.D. Mermin, Solid State Physics. Saunders College, Philadelphia **116**, 217 (1976)
2. P. Giannozzi, S. Baroni, N. Bonini, M. Calandra, R. Car, C. Cavazzoni, A. Dal Corso, Quantum ESPRESSO: modular and open-source software Project for quantum simulations of materials. J. Phys. Condens. Matter **21**(39), 395502 (2009)
3. J. Hafner, Ab-initio simulations of materials using VASP: Density-functional theory and beyond. J. Comput. Chem. **29**(13), 2044–2078 (2008)
4. J.P. Hirth, A brief history of dislocation theory. Metall. Trans. A **16**, 2085–2090 (1985)
5. P. Hohenberg, W. Kohn, Inhomogeneous electron gas. Phys. Rev. **136**(3B), B864 (1964)
6. Kittel, C. (1953). Introduction to solid state physics (1st ed.) John Wiley & Sons.
7. Kostorz, G. (1987). Dislocations in solids edited by H. Suzuki, T. Ninomiya, K. Sumino and S. Takeuchi. *Acta Crystallographica Section A: Foundations of Crystallography*, *43*(4), 592–592.
8. M. Nepal, G. Paudel, S. Aryal, A. Devkota, H.K. Neupane, Adsorption of Water on Vacancy Defective h-BN Bilayer at B and N Sites: First-principles calculation. BIBECHANA **21**(2), 129–141 (2024)
9. H.K. Neupane, N.P. Adhikari, Structure, electronic and magnetic properties of 2D Graphene-Molybdenum diSulphide (G-MoS2) Heterostructure (HS) with vacancy defects at Mo sites. Computational Condensed Matter **24**(2020), e00489 (2020)
10. H.K. Neupane, N.P. Adhikari, First-Principles Study of Defected Single Layer Hexagonal Boron-Nitride (h-BN). Nepal Journal of Science and Technology **20**(2), 55–66 (2021)
11. H.K. Neupane, N.P. Adhikari, First-principles study of structure, electronic, and magnetic properties of C sites vacancy defects in water adsorbed graphene/MoS2 van der Waals heterostructures. J. Mol. Model. **27**(3), 82 (2021)
12. Neupane, H. K., & Adhikari, N. P. (2021). Effect of vacancy defects in 2D vdW graphene/h-BN heterostructure: First-principles study. *AIP Advances*, *11*(8).
13. H.K. Neupane, B. Bhattarai, N.P. Adhikari, Transport properties of B and N sites vacancy defects graphene/h-BN heterostructures: First-principles study. Physica B **666**, 415077 (2023)
14. H. Neupane, P. Khatri, A. Devkota, N.P. Adhikari, Electronic and magnetic performance of MoS2monolayer via Tc & Nb impurities defect and water adsorption. BIBECHANA **20**(3), 224–235 (2023)
15. Rijal, O. S., Neupane, H. K., Oli, D., Neupane, R. K., Shrestha, P., Sharma, S., & Parajuli, R. (2025). A First-Principles Investigation of the Structural, Mechanical, Dynamic, Electronic, Magnetic and Optical Properties of Ti2AC (A= Cd, S) MAX Phase Compounds. *Journal of Physics D: Applied Physics*.).

16. P. Rodriguez, Sixty years of dislocations. Bull. Mater. Sci. **19**, 857–872 (1996)
17. Sturdevant, C. (2001). BMC widens its Win2K Patrol. *eWeek*, 61–61.

Chapter 10
Beyond DFT Methods

Shivani Bhardwaj and Sudhir K. Pandey

Abstract Density functional theory (DFT) has provided a practical approach to calculate and predict the electronic structure of atoms, molecules and solids. It has been known to have revolutionized the computational quantum mechanics by offering a remarkable balance between the computational efficiency and predictive power by reducing the many-body problem to one with effective single-particle periodic potential. However, its applicability is limited by its inability to capture the dynamics and many-body effects in materials displaying strong electronic-correlation effects. Subsequently, due to its static nature, it is found to fail in describing the observed electronic excitation properties of solids. These broad limitations have necessitated to resort to methods that extend beyond DFT. Some of the methods like time-dependent density functional theory (TDDFT), DFT+U, GW, and Hybrid functionals were developed and have been known to explain numerous physical properties of wide variety of solids. There are few other methods such as DFT+DMFT, GW+DMFT, which belong to recent theoretical advancements in the direction of beyond DFT methods. However, we will restrict our discussion to TDDFT, DFT+U, GW, and Hybrid functionals.

10.1 Time Dependent Density Functional Theory

Hohenberg and Kohn (HK) in 1964 [1] proposed that with the knowledge of its ground-state density, a stationary electronic system can be fully described. And in principle, from this ground-state density all the static physical observables as well as the interacting many-body wave function can be determined. Consideration of the

S. Bhardwaj (✉)
School of Physical Sciences, Indian Institute of Technology Mandi, Kamand, India
e-mail: sbhardwajiitm369@gmail.com

S. K. Pandey
School of Mechanical and Materials Engineering, Indian Institute of Technology Mandi, Kamand, India
e-mail: sudhir@iitmandi.ac.in

R. Thotagamuge (ed.), *Density Functional Theory*,
https://doi.org/10.1007/978-981-95-9745-1_10

charge density proved to be a convenient variable, as the degrees of freedom were reduced from $3N$ spatial coordinates to just three spatial coordinates, which is evident from the proof of HK theorem based on the Rayleigh-Ritz principle. The major breakthrough of was HK's theorem of obtaining the accurate ground-state density of interacting many-body system with an auxiliary non-interacting system. The ground-state energy which is a ground-state density functional can thus be obtained through variational principle in terms of charge density. The minimisation of this total energy with respect to charge-density leads to the lowest energy state or the non-degenerate ground-state of the system).

The consideration of charge density as the fundamental variable and a non-interacting auxiliary system (or Kohn-Sham system) which yields the exact charge density for the interacting system sets the basis for the modern day density functional theory (DFT) [2]. In its standard form, DFT has grown up to be a method of choice for calculating the ground state observables for a large number of systems by replacing the many-body (many-electron) problem with an effective single-electron problem which can be solved exactly. Within this, a single particle Schrodinger equation is solved for the system in an effective external potential, v_{ks}, which is also the functional of density ($\rho(\mathbf{r})$). The effective external potential can be usually written in different contributing potentials, i.e.,

$$v_{ks}[\rho](\mathbf{r}) = v_{ext}[\rho](\mathbf{r}) + v_H[\rho](\mathbf{r}) + v_{xc}[\rho](\mathbf{r}) \tag{10.1}$$

Here, v_{ext} (external potential) refers to the general Coulomb interaction between electrons and nuclei. v_H is the Hartree potential, which represents the traditional electron-electron interaction term, also known as direct Coulomb interaction term. The v_{xc} potential term contains the many-body interaction effects and is the unknown term in the DFT Hamiltonian equation. There have been approximations developed to resort to, for the exchange-correlation potential of the systems for example, Local (spin) density approximation (L(S)DA), Generalized gradient approximation (GGA). Despite its enormous success, DFT suffers from limitations. The static nature of DFT and $\rho(\mathbf{r})$ fail miserably to describe the excitation properties of real solids. In other words, account of the dynamical response of a system in the presence of time dependent perturbation cannot be understood with DFT. For instance, the use of laser pulse in an optical experiment or any generic transverse electromagnetic wave for that matter (within the optical limit) can serve as the time dependent external perturbation (δv), which can be expressed as Eq. 10.2 in dipole approximation.

$$\delta v(\mathbf{r}, t) = E_0 f(t) \sin(\omega t)\mathbf{P} \cdot \mathbf{r} \tag{10.2}$$

Here, E_0, ω, and $\mathbf{P}$ are the amplitude, frequency, and polarization of the laser pulse. The $f(t)$ is the pulse envelope. These limitations of DFT were promised to be tackled by time dependent DFT (TDDFT), which is the time dependent extension of DFT. The foundation of TDDFT was laid by Runge and Gross (RG) in 1984 [3], where they proposed a HK like theorem by solving time-dependent Schrodinger equation for time dependent systems. The original motive of the Runge and Gross included studying the time dependent scattering experiments in systems. However, with the advent of TDDFT the study of various excitation spectra of systems, for example, photoemission spectra, inverse-photoemission spectra, optical spectra was also promised.

The main thesis of RG theorem shows that the time-dependent charge density ($\rho(\mathbf{r}$,t)) would be sufficient to solve the single particle time-dependent Schrodinger equation for time-dependent systems. In consequence, parallel to DFT, RG prove the direct correspondence between the time-dependent external potential $v_{ext}(\mathbf{r}, t)$ and charge density $\rho(\mathbf{r}, t)$ for many-body systems evolving from a fixed initial state. Therefore, TDDFT can be seen as an extension of the HK DFT theorem to the time-dependent system. In analogy to DFT, the time-dependent Schrodinger equations can be solved exactly by constructing an auxiliary non-interacting system to obtain the real density of the system. Subsequently, all other physical observables can be obtained from the solutions of time-dependent independent particle DFT equations. This is very profound statement as it says that all other properties of the quantum systems can be obtained once we know its charge density. Analogous to static DFT where the exchange correlation potential is to be approximated, here also in TDDFT the time dependent exchange correlation potential is unknown and has to be approximated to solve the single particle time-dependent Schrodinger equation.

$$-i\hbar\frac{\delta}{\delta t}\psi(\mathbf{r}_1,.....\mathbf{r}_N,t) = H(\mathbf{r}_1,.....\mathbf{r}_N,t)\psi(\mathbf{r}_1,.....\mathbf{r}_N,t) \tag{10.3}$$

$$\rho(\mathbf{r},t) = \int \psi(\mathbf{r}_1,.....\mathbf{r}_N,t)\cdot\psi^{\dagger}(\mathbf{r}_1,.....\mathbf{r}_N,t)\delta\mathbf{r}_1...\delta\mathbf{r}_N \tag{10.4}$$

However, we will understand the central RG theorem which laid the foundation of TDDFT.

The Runge Gross Theorem: The discussion till now concludes that in case of static DFT, the ground state density ($\rho(\mathbf{r})$) and external potential ($v_{ext}(\mathbf{r})$) have one to one correspondence, i.e., F: $v_{ext}\mathbf{r}) \Longrightarrow \rho(\mathbf{r})$ is invertible. This is fine if we are interested in the ground state of the system, but is insufficient if we are interested in the excitation energies or response of the system to an external time dependent perturbation of any form, where the external potential and densities are no more static and thus DFT does not apply. Interestingly, RG came up with their theorem showing the one to one correspondence between time-dependent external potential ($v_{ext}(\mathbf{r}, t)$) and charge density ($\rho(\mathbf{r}, t)$). The main theorem of RG's formulation states that for every single particle time-dependent external potential ($v_{ext}(\mathbf{r}$,t)), which is Taylor series expandable in time around $t = t_0$, a map established as F:$v_{ext}(\mathbf{r}$,t) $\rightarrow \rho(\mathbf{r}$,t) can be defined from the time-dependent Schrodinger equation with a fixed initial state $\psi(t_0) = \psi_0$, where ψ_0 is the ground state density and obtaining correspondingly the charge densities. This mapping in F is invertible merely up to an additive time dependent function in the potential. The theorem actually proves the invertibility of F for time-dependent case ($v_{ext}(\mathbf{r}, t)$ and $\rho(\mathbf{r}, t)$), by assuming two external potentials

($v_{ext}(\mathbf{r}, t)$ and $v'_{ext}(\mathbf{r}, t)$) which differ by more than a time-dependent scalar function (such that they follow Eq. 10.5).

$$v_{\text{ext}}(\mathbf{r}, t) - v'_{\text{ext}}(\mathbf{r}, t) \neq c(t) \tag{10.5}$$

This is important since if the potentials only differ by a scalar time-dependent function then they will yield same charge density (charge density will no longer be unique). This is because the wavefunctions corresponding to these two potentials would differ only by a phase difference, i.e., -iαt and which would get cancelled while calculating the charge density or in fact any observable for that matter (ψ(t) $\rightarrow \psi'$(t) = $e^{-i\alpha t}\psi$(t)), hence yielding the same charge density. This issue would not let the F to be invertible for the time-dependent case. This was however overcome by showing that no two time-dependent potentials differing by more than a scalar time-dependent function result into the same charge density. It is also shown however that the observables do not depend on this time dependent phase factor of the wavefunction. This is illustrated by showing that the current densities $\mathbf{j}$ and $\mathbf{j}'$ produced by $v_{ext}(\mathbf{r}, t)$ and $v'_{ext}(\mathbf{r}, t)$ are different. It is further demonstrated that if two systems have different current densities, then they must also possess different time-dependent densities, i.e., if $\mathbf{j} \neq \mathbf{j}' \implies \rho(\mathbf{r}, t) \neq \rho(\mathbf{r}, t)$. This proof is shown with help of set of hydrodynamical equations or continuity equations. Conclusively, once the external potential and initial state wavefunction is fixed, every observable O can be determined which stresses the underlying idea that every observable is a unique functional of $\rho(\mathbf{r}, t)$.

$$O(t) = O[\rho, \psi_0](t) \tag{10.6}$$

It can be recalled that there exists variational principle in static DFT with respect to charge density, wherein the energy minina is found at the exact ground-state charge density. Here, in TDDFT, since energy is not conserved in the time-dependent systems, there's no variational principle. However, there is an action integral formulation, where the stationary points of the action correspond to the exact time-dependent charge density. This also says that the exact time-dependent charge density can be found by varying this action and finding the stationary points. It results into a fairly convenient formulation where a time-dependent Kohn-Sham like non-interacting system can be constructed. Now suppose an auxiliary non-interacting system of electrons with external potential v_{ks}. As per the RG theorem this potential is unique, giving a density which is the density of original interacting system. We can now write the time-dependent Schrodinger equation for the system as

$$-i\hbar\frac{\delta}{\delta t}\psi(\mathbf{r}, t) = \hat{H}_{ks}(\mathbf{r}, t)\psi(\mathbf{r}, t) \tag{10.7}$$

And the density of the original interacting system can be determined from the Kohn-Sham orbitals as

$$\rho(\mathbf{r}, t) = \sum_i |\phi_i(\mathbf{r}, t)|^2 \tag{10.8}$$

where Kohn-Sham Hamiltonian can be defined as:

$$\hat{H}_{ks}(\mathbf{r}, t) = -\frac{\nabla^2}{2m} + v_{ks}[\rho](\mathbf{r}, t) \tag{10.9}$$

Similar to the Kohn-Sham ground state DFT, in TDDFT, the $v_{ks}[\rho](\mathbf{r}, t)$ can be decomposed into the time dependent static counterparts.

$$v_{ks}[\rho](\mathbf{r}, t) = v_{ext}[\rho](\mathbf{r}, t) + v_H[\rho](\mathbf{r}, t) + v_{xc}[\rho](\mathbf{r}, t) \tag{10.10}$$

The exchange-correlation potential $v_{xc}(\mathbf{r}, t)$ carries the many-body effects information and its form of functional dependence on density is essentially unknown. It is non-local potential in space and time, which implies that its value at any point in time t and space $\mathbf{r}$ depends on the density at all other positions and all previous times (causality principle). The exchange-correlation potential term is approximated using different physical and mathematical formulations. The results of TDDFT calculations largely depend on the accuracy of approximations to this exchange-correlation potential.

Whenever time-dependent perturbation is small relative to the external static potential, the perturbation theory can be used on the top of DFT, i.e., Density functional perturbation theory (DFPT). However, when this perturbation is large enough and the system experiences a small departure from the ground state equilibrium, the TDDFT can be used in the linear response regime from the static DFT. This situation mostly arises while studying the energy-loss spectra or X-ray absorption and inelastic X-ray scattering spectroscopy, or optical absorption spectroscopy. The TDDFT has been immensely successful in providing the excitation energies and spectra in these aforementioned cases. The advancements in laser technology with more intense laser probes and possible study of non-equilibrium system dynamics has led to promising applications of TDDFT technique beyond linear response range. The theory faces challenges in developing approximations for the suitability in non-perturbative regime; for instance being able to include the electron-ion dynamics in order to study the photo-explosion or photo-chemistry. Moreover, TDDFT has also been successfully applied to organic molecules for the UV-Vis absorption spectra [4]. It is also indicated across studies that TDDFT can handle large systems sizes as opposed to the wave-function based methods thus enabling exploration of complex systems such as biomolecules [5]. This shows its computational efficiency. Also, due to the advent of various exchange-correlation functionals within TDDFT, it is now becoming easy to tailor system specific calculations leading to enhanced accessibility of the method across different chemical contexts. TDDFT has also been applied to study the optical properties of nanostructures, which is crucial to designing the materials for photovoltaic applications [6]. Moreover, TDDFT also finds relevance in exciton dynamics and non-adiabatic processes, where real-time TDDFT has been employed to model charge transfer dynamics in photovoltaic materials, thus providing insights into the efficiency of energy conversion processed. However, the theory suffers from certain failures or limitations and invites developments in various domains. For example, the standard exchange-functionals often, for the description of charge-transfer excitations, leading to large errors in predicted excitation energies for the systems wherein such transitions are prominent [7]. The theory also fails significantly in case of strongly correlated electron systems due to strong correlation effects resulting in inaccurate electronic excitation and magnetic properties. It is often seen to be inaccurate in case of double excitations and also in the study of plasmonic systems where it can predict spurious resonances [8]. Despite these challenges, TDDFT has been extensively successful and powerful tool in electronic structure studies as it is both computationally efficient and accurate for a variety of applications. However, its limitations, in particular concerning the treatment of multi-excitations, strongly correlated electron systems, provide the scope for further developments and need further consideration. Ongoing advancements into the development of improved exchange-correlation functionals and other such hybrid methods aim to address these challenges and are expected to enhance the applicability of TDDFT in complex electronic structure problems.

DFT+*U* Method

Density functional theory (DFT) has been a widely used computational method for decades to study the electronic structure properties of a wide variety of materials. The DFT equations use approximations to the exchange-correlational energy using variety of functionals such as Local (spin) density approximation (L(S)DA) [9] or Generalized gradient approximation (GGA). Using these functionals DFT has been able to successfully derive numerous electronic structure properties of materials. However, these functionals have been found often inadequate in describing the observed properties of strongly correlated electron systems as inferred from the earliest of studies carried out on late transition metal oxides. Later it became evident that these functionals or DFT fails in understanding the fundamental properties of strongly correlated electron systems which include systems with open d or f shells. For instance, since the functionals, LDA, and GGA in DFT are essentially orbital-independent potentials, thus the application of DFT to the strongly correlated electron systems with partially filled d or f bands results into a metallic solution. However, this could not be the solution for late-transition metal oxides or rare-earth metal compounds where these narrow band electrons are well localized and inhabit significant energy difference between the occupied and unoccupied orbitals/subbands of the same band (which are also distinguished as lower and upper Hubbard bands in a model Hamiltonian approach) and should give a band gap as opposed to DFT obtained metallic solution. Following the issues in describing the strongly correlated electron systems, there have been attempts to improve on the existing DFT functionals to account for the observed properties of these systems. The primary approach employed was identified to be self-interaction correction (SIC) [10] and inaccurate account of the strong on-site Coulomb interaction within these localized orbitals. Althought the SIC was able to account for the localized nature of the d and f electrons in these systems, but the single-particle energies were estimated quite off from the observed energies in the spectroscopic experiments. Later, a Hubbard correction term was employed to treat these localized electrons on the top of DFT. It was initially successfully formulated for NiO to show the band gap opening consistent with the experimental findings. The insulating behavior of NiO was then confirmed theoretically using DFT+U approach. The original DFT+U or LDA+U approach was developed in 1991 by Anisimov, Zaanen, and Andersen [11]. This approach was then widely utilized to improve on the description of electron localization by augmenting the standard DFT with a Hubbard U correction [12].

The formulation is done to treat the localized d or f electrons with Hubbard U and the rest itinerant electrons (electrons belonging to s, p shells) with the DFT functionals, i.e., LDA or GGA. This can be understood as dividing the system into two subsystems (i) one comprised of d or f electrons for which the electron-electron (e-e) Coulomb interaction should be taken into account with Hubbard U (representative of the strength of e-e interaction) (ii) second where the itinerant s-, p-electrons are approximated by the orbital-independent single-electron potential (LDA or GGA). To avoid the double counting or double consideration of the localized electrons, a double counting term E_{dc}, is subtracted (for the d or f electrons; since they have been considered twice: once within the DFT functional and again with the U treatment) from the total energy, E_{DFT+U} expression given as follows [13].

$$E_{DFT+U} = E_{DFT} + E_U - E_{dc} \tag{10.11}$$

The e-e Coulomb interaction energy (E_U) is represented as $\frac{1}{2}U\sum_{i\neq j} n_i n_j$ (where n_i are the d- or f-orbital occupancies) in the Hamiltonian of these systems.

Assuming in DFT the e-e Coulomb interaction energy can be written in terms of total number of electrons in the shell we get

$$E = UN(N-1)/2 \tag{10.12}$$

where $N = \sum_i n_i$; It is essential to note that total energy of the d or f shells is accurate to a good approximation within DFT (LDA or GGA), not the orbital eigenvalues. This term (Eq. 10.12) when subtracted from the E_{DFT} and then adding the Hubbard term would result in what is famously called DFT+U functional, given by as follows:

$$E_{DFT+U} = E_{DFT} - UN(N-1)/2 + \frac{1}{2}U\sum_{i\neq j} n_i n_j \tag{10.13}$$

Here, for convenience we have omitted the exchange interaction term (J) from the expression. This is for demonstrating the idea how DFT+U can give insulating ground state. However, one needs to include J along with U to do realistic calculations. The orbital energies (ϵ_i) or the orbital eigenvalues can be derived from the derivative of Eq. 10.13 with respect to the orbital occupancies n_i, i.e.,

$$\epsilon_i = \frac{\delta E_{DFT+U}}{\delta n_i} = \epsilon_{DFT} + U(\frac{1}{2} - n_i) \tag{10.14}$$

Note that based on the n_i the DFT orbital energy (ϵ_{DFT}) can shift by $+U/2$ (for $n_i = 0$; for unoccupied orbitals) or $+U/2$ (for $n_i = 1$; for occupied orbitals). And similarly the orbital-dependent potential $V_i(\mathbf{r})$ can be defined as

$$V_i(\mathbf{r}) = \frac{\delta E}{\delta \rho_i(\mathbf{r}))} \tag{10.15}$$

It should be noted that here $\rho_i(\mathbf{r})$ is the orbital charge density as opposed to the total charge density considered in the DFT framework. Note that this orbital dependent potential clearly shows that the highest occupied orbital subband (lower Hubbard band) and lowest unoccupied subband (upper Hubbard band) are separated by an energy equal to U (strength of e-e Coulomb interaction). This reproduces the band gap in transition metal oxides which are thus identified to be Mott-Hubbard insulators. In the presence of spin polarization the Coulomb interaction term can be written as (considering the atomic nature is retained and thus expanding in the localized orthonormal basis):

$$E^U[\{n\}] = \frac{1}{2}\sum_{\{m\},\sigma}\Big\{\langle m, m''|V_{ee}|m', m'''\rangle n^{\sigma}_{mm''} n^{-\sigma}_{m'm'''} + \Big(\langle m, m''|V_{ee}|m', m'''\rangle - \langle m, m''|V_{ee}|m', m''\rangle\Big) n^{\sigma}_{mm''} n^{\sigma}_{m'm'''}\Big\} \tag{10.16}$$

where $n^{\sigma}_{m'm''}$ are the density matrix elements for nl state, where m is the magnetic quantum number, n is the principle quantum number, l is the angular quantum number, given by

$$n^{\sigma}_{m'm''} = -\frac{1}{\pi}\int^{E_F} ImG^{\sigma}_{inlm',inlm''}(E)dE \tag{10.17}$$

Here $G^{\sigma}_{inlm',inlm''}(E) = \langle inlm'\sigma|(E - \hat{H})^{-1}|inlm''\sigma\rangle$ are matrix elements of Green function.

The U_{ee} are the screened Coulomb interactions between the electrons of nl states. E_{dc} can be written as

$$E_{dc}[n^{\sigma}] = \frac{1}{2}UN(N-1) - \frac{1}{2}J[N^{\uparrow}(N^{\uparrow}-1) + N^{\downarrow}(N^{\downarrow}-1)] \tag{10.18}$$

Here, N is the total number of up- and down- spin electrons, $N^{\uparrow}$ and $N^{\downarrow}$ refer to the number of up- and down-spin electrons, respectively. U is the screened Coulomb interaction parameter and J is the exchange interaction parameter [13, 14].

There are two extensively developed schemes in treating the double-counting term in the DFT+U method namely, Around mean field (AMF) [15] and Fully localized scheme (FLL) [16]. The AMF scheme is found to be suitable for systems with weaker electronic correlations and exhibit significant overlap between the itinerant and localized states. On the other hand, FLL scheme generally works well for strongly correlated electron systems such as late transition metal oxides.

$$\hat{H} = \hat{H}_{\mathrm{DFT}} + \sum_{mm'} |inlm\sigma\rangle V^{\sigma}_{mm'} \langle inlm'\sigma|, \tag{10.19}$$

$$V^{\sigma}_{mm'} = \sum_{m''} \Big\{ \langle m, m''|V_{ee}|m', m'''\rangle n^{-\sigma}_{m'm'''} + \Big(\langle m, m'|V_{ee}|m', m'''\rangle - \langle m, m''|V_{ee}|m', m'\rangle \Big) n^{\sigma}_{m'm'''} \Big\} \tag{10.20}$$

$$- U\Big(N - \frac{1}{2}\Big) + J\Big(N^{\sigma} - \frac{1}{2}\Big) \tag{10.21}$$

The total DFT+U Hamiltonian can thus be represented as Eq. 10.19, in the presence of spin-polarization. The implementation of DFT+U methodology often considers the effective Coulomb interaction parameter $U_{eff} = U$-J. This in a way simplifies the account of correlations by considering the competition between the Coulomb repulsion and exchange effects. It also reduces the number of parameters and as a consequence leads to improved computational efficiency with accuracy. This in particular can be used in case of systems for which J is significantly small in comparison to U [17].

DFT+U has been successfully used to study the correlated electron systems which are systems with open d or f shells. In materials transition metal oxides like FeO, NiO and CeO_2, the band gap has been accurately produced. It is also able to describe the magnetic ordering including antiferromagnetic states in oxides for example MnO, CoO. Numerous structural properties such as lattice constants, bond lengths, have been correctly predicted by DFT+U in comparison to DFT. However, the DFT+U functional suffers from several limitations. The choice of U is one of the difficult tasks. Finding material specific U has been a tedious task since its use. The use of inaccurate value of U can lead to nonphysical results for the materials' properties. Furthermore, in systems exhibiting mixed character (localized and itinerant electrons), DFT+U fails to provide a balanced description. In spite of being able to provide an improved picture for strongly correlated electron systems, it still lacks in better quantification and several properties where dynamical correlation effects are critically relevant since DFT+U assumes a static mean-field treatment of correlations. Further, DFT+U method shows sensitivity to the starting correlated orbital occupancy distribution or electronic configuration of the system which can give rise to metastable state solution rather than the true ground state solution. Therefore, DFT+U methodology is prone to get stuck in local energy minima and hence can miss out the global energy minima or the lowest ground state leading to misinterpretation of ground state properties of the system. The account of long-range correlation effects fails in DFT+U which can captured by more advanced theoretical methods like GW [18].

GW

GW approximation (GWA) has become immensely powerful method in understanding the electronic structure of materials, because it can directly determine the quasiparticle excitations. GWA was formulated based on many-body perturbation theory. GWA has succeeded in addressing the electronic band gaps and excitation spectra. The "G" stands for Green's function and "W" represents the screened Coulomb interaction. Green's function describes the propagation of an electron or hole in the material and screened Coulomb interaction accounts for the effect of electronic correlations in the material.

The mean-field theories fail to give the quasiparticle energies for the many-body systems. The corrections to mean-field theories have been able to improve the total energies but do not succeed in describing the quasiparticle energies. Also, the band gap problem in DFT stays an issue even with the mean-filed corrections like SIC or with the use of GGA exchange-correlation functional. Moreover, DFT+U is designed for strongly correlated systems where the degeneracy in localized states is split by the Coulomb interaction parameter into the upper and lower Hubbard band. The KS construction of DFT misses the energy dependence of the screened exchange-correlation functional, which can be very important in certain cases. The KS scheme also suffers from the choice of keeping the exchange-correlation potential non-local. It might provide an inadequate picture in cases where long-range interactions cannot be ignored. As mentioned before, GWA is derived from many-body perturbation theory. The self-energy used in GWA includes a dynamically screened Coulomb interaction. The self-energy is non-local and energy dependent.

The self-energy operator $\Sigma(\mathbf{r}, \mathbf{r'}, \omega)$ encapsulates the many-body interactions beyond the mean-field approximation. The self-energy operator is expressed as

$$\Sigma(\mathbf{r}, \mathbf{r}', \omega) = \frac{i}{2\pi} \int d\omega' G(\mathbf{r}, \mathbf{r}', \omega - \omega') W(\mathbf{r}, \mathbf{r}', \omega') \tag{10.22}$$

The screened Coulomb interaction (W) captures the screening effect in the material with dielectric constant $\varepsilon(\mathbf{r}, \mathbf{r}'', \omega)$ and gives a measure of the strength of response of the system to an external perturbation is expressed as below

$$W(\mathbf{r}, \mathbf{r}', \omega) = \int d^3\mathbf{r}'' \varepsilon^{-1}(\mathbf{r}, \mathbf{r}'', \omega) v(\mathbf{r}'', \mathbf{r}') \tag{10.23}$$

The dielectric constant is often estimated using Random Phase Approximation (RPA). The GWA provides the quasiparticle energies via replacing the exchange-correlation potential in DFT or any mean-field theory by self-energy. The earliest attempt to go beyond the mean-field approximations in the form of GW was made by Quinn and Ferrell in 1958 for the electron gas [19]. Later on various implementations came in until 1985, it widely came to be known as GWA which can be utilized for a variety of materials.

Generally, to study the electronic excitation spectrum of solid, the photoemission experiments are done. In such experiment photons with certain energy ω are used to knock out the electrons which will be then called photoelectrons. The kinetic energy (K.E) of photoelectrons are measured along a direction **k** and the excitation spectrum is calculated.

$$\omega = K.E + B.E(\mathbf{k}) \tag{10.24}$$

where $B.E$ is the binding energy of the photoelectrons. The photoemission experiment measures the excitation spectrum of a solid with the presence of a hole created in the process (from occupied density of states). Similarly, we can get the excitation spectrum for the unoccupied states using inverse photoemission spectrum. In the limit of large K.E of photoelectron (which is commonly known as sudden approximation), the spectrum can be directly related to the single-particle Green's function defined as:

$$iG(x, x') = \langle N|T[\hat{\psi}(x)\hat{\psi}^{\dagger}(x')]|N\rangle \tag{10.25}$$

$$\begin{cases} \langle N|\hat{\psi}(x)\hat{\psi}^{\dagger}(x')|N\rangle & \text{for } t > t' \text{ (electron)} \\ -\langle N|\hat{\psi}^{\dagger}(x')\hat{\psi}(x)|N\rangle & \text{for } t < t' \text{ (hole)} \end{cases} \tag{10.26}$$

where $\hat{\psi}(x)(\hat{\psi}(x)^{\dagger})$ is the annihilation (creation) operator for an electron at $x = (\mathbf{r}, t)$ and field operator, $|N\rangle$ is the exact N-electron ground state and T is the time-ordering operator. The physical interpretation of Green's function can be associated with photoemission and inverse photoemission processes. For instance, Green's function for $t' > t$, correspond to the probability amplitude that a hole created at x will propagate to x' and similarly, for $t > t'$, the probability amplitude that an electron added at x' will propagate to x. Thus, the Green function describes the photoemission and inverse photoemission processes.

The single particle spectral function can be expressed directly in terms of Green's function as follows:

$$A(\mathbf{r}, \mathbf{r}', \omega) = -\frac{1}{\pi} ImG(\mathbf{r}, , \mathbf{r}', \omega) \tag{10.27}$$

Also, from Green's function, the expectation value of any single-particle operator in the ground state can be calculated along with the ground-state energy and single-particle excitation spectrum.

GWA captures quasiparticle excitations, making it useful in studying electronic excitation spectra in materials. GWA has helped in predicting optical and electronic properties of wide variety of materials ranging from semiconductors to metals. This theory also been able to describe the band structure of graphene and transition metal dichalcogenides. Interestingly, GWA can be used to serve as a benchmark for more advanced methods like hybrid functionals and DFT+U in validating their predictions. Despite, the theory suffers from certain limitations. This method is computationally expensive due to evaluation of dielectric matrix and Green's function. High computational demands restrict its application to small or medium sized systems, limiting its use in complex materials [20]. GW is often performed as a "one-shot" correction, which relies on initial DFT wavefunctions and eigenvalues. The results of GW calculations are sensitive to technical parameters such as the number of bands, point sampling and the dielectric matrix cut off. Moreover, achieving convergence can be computationally prohibitive, especially for systems with a large number electrons. GWA does not fully capture strong electron-electron correlation effects, such as those in Mott insulators or systems with localized electrons. GWA works better in treating long range correlations but systems where short-range correlations are significant fail to be described by GWA. In spite of these limitations, ongoing advancements in computational algorithms, hybrid methods such as GW+DMFT continue to expand the reach of GWA.

Hybrid Functionals

DFT has been known to revolutionize the computational chemistry and material science since its development [21]. The approximations to exchange-correlation functional have been since then the center of improvement to describe the properties of varieties of materials. Among various exchange-correlation functionals developed within DFT, hybrid functionals stand out for their ability to enhance predictive accuracy by incorporating a portion of exact exchange from Hartree-Fock theory. In 1933, Axel Becke introduced the hybrid approach of constructing the density functional approximation [22].

Hybrid functionals can be seen as a class of exchange-correlation functionals that mix the exchange energy calculated from Hartree-Fock (HF) theory with exchange and correlation energies derived from DFT approximations.

Mathematically, hybrid functionals can be expressed as

$$E_{xc}^{hybrid} = cE_x^{HF} + (1-c)E_x^{DFT} + E_c^{DFT} \tag{10.28}$$

Here, c is the mixing coefficient which can be often empirically determined through theoretical considerations. E_x^{HF} is the exact exchange-energy from Hartree-Fock theory, E_x^{DFT} is the exchange-energy from a DFT functional, E_c^{DFT} is the correlation energy from a DFT functional.

B3LYP the most common hybrid functional made from combining the Becke exchange functional ($E_x^{Becke88}$), three parameter exchange functional (B3) [23] with Lee-Yang-Parr correlation functional (LYP) (E_c^{LYP})[24] and E_c^{LYP} is the VWN local spin density approximation to the correlation functional [25]. It is typically expressed as:

$$E_{xc}^{B3LYP} = a_0 E_x^{HF} + a_x E_x^{Becke88} + (1-a_x)E_x^{LDA} + a_c E_c^{LYP} + (1-a_c)E_c^{VWN} \tag{10.29}$$

where a_0, a_x, and a_c are empirically determined coefficients. It is often found that pure DFT functionals like LDA and GGA often underestimate or overestimate the exchange-correlation energy. Whereas, the hybrid functionals by blending a portion of Hartree-Fock exact exchange, correct for these errors and improve the description of exchange and correlation. Hybrid functional have been able to provide accurate estimation of the band gaps in solids. Apart from this Hybrid functionals are able to better capture the reaction barriers compared to pure DFT methods. They provide more reliable excitation energies, improving predictions for spectroscopy. There is issue of self-interaction error with pure DFT since the functional doesn't cancel out the spurious self interaction of electron, however, the hybrid functionals are able to mitigate this by including Hartree Fock exchange thus making it self-interaction free.

Some of the popular examples of hybrid functionals include

B3LYP, PBE0 [26], HSE [27] where PBE0 combines the Perdew-Burke-Ernzerhof(PBE) GGA functional with Hartree-Fock exchange. HSE is the screened hybrid functional for condensed-matter systems.

Few of the notable successes of hybrid functionals include the following: B3LYP [28] and PBE0, hybrid functionals have demonstrated success in predicting reaction energies [29], electron affinities [30], ionization potentials [31], bond-lengths, bond-angles and other geometrical molecular aspects such as vibrational frequencies [32]. HSE06 is widely identified for accurate band structure calculations in solids [33], thus providing a bridge between DFT and many-body methods like GW [34].

References

1. P. Hohenberg, W. Kohn, Phys. Rev. B **136** (1964)
2. W. Kohn, L. Sham, J. Phys. Rev. A **140** (1965)
3. E. Runge, E.K.U. Gross, Phys. Rev. Lett. **52** (1984)
4. A.D. Laurent, D. Jacquemin, Int. J. Quantum Chem. **113** (2013)
5. D.F. Tegegn, H.Z. Belachew, A.O. Salau, Sci. Rep. **8**, 8146 (2024)
6. M.W.D. Hanson-Heine, M.W. George, N.A. Besley, Mol. Phys. **11**, 1452 (2018)
7. A. Prlj, B.F.E. Curchod, A. Fabrizio, L. Floryan, C. Corminboeuf, J. Phys. Chem. Lett. **6**, (2015)
8. A.A.B. Baloch, M.I. Hossain, N. Tabet, F.H. Alharbi, J. Phys. Chem. Lett. **6**, 1 (2015)
9. P. Hohenberg, W. Kohn, Phys. Rev. **136**, B864 (1964); W. Kohn, L.J. Sham, ibid. **140**, A1133 (1965)
10. A. Svane, O. Gunnarsson, Phys. Rev. Lett. **65**, 1148 (1990)
11. V.I. Anisimov, J. Zaanen, O.K. Andersen, Phys. Rev. B **44**, 943 (1991)
12. V.I. Anisimov, I.V. Solovyev, M.A. Korotin, M.T. Czyzyk, G.A. Sawatzky, Phys. Rev. B **48**, 16929 (1993)
13. O. Gunnarsson, O.K. Andersen, O. Jepsen, J. Zaanen, Phys. Rev. B **39**, 1708 (1989)
14. V.I. Anisimov, O. Gunnarson, Phys. Rev. B. **43**, 7570 (1991)
15. M.T. Czyżyk, G.A. Sawatzky, Phys. Rev. B. **49** (1994)
16. V.I. Anisimov, J. Zaanen, O.K. Andersen, Phys. Rev. B **44**, 943 (1991)
17. S.L. Dudarev, G.A. Botton, S.Y. Savrasov, C.J. Humphreys, A.P. Sutton, Phys. Rev. B **57**, 1505 (1998)
18. M. Cococcioni, S. de Gironcoli, Phys. Rev. B **71**, 035105 (2005)
19. R.A. Ferrell, J.J. Quinn, Phys. Rev. **112**, 812 (1958)
20. L. Reining, WIREs Comput. Mol. Sci. **8**, e1344 (2018)
21. R.M. Martin, *Electronic Structure: Basic Theory and Practical Methods* (Cambridge University Press, 2004)

22. A.D. Becke, J. Chem. Phys. **98**(2), 1372 (1993)
23. A.D. Becke, Phys. Rev. A **38**(6), 3098 (1988)
24. C. Lee, W. Yang, R.G. Parr, Phys. Rev. B. **37**(2), 785 (1988)
25. S.H. Vosko, L. Wilk, M. Nusair, Can. J. Phys. **58**(8), 1200 (1980)
26. J.P. Perdew, M. Ernzerhof, K. Burke, J. Chem. Phys. **105**(22), 9982 (1996)
27. J. Heyd, G.E. Scuseria, M. Ernzerhof, J. Chem. Phys. **118**(18), 8207 (2003)
28. A.D. Becke, J. Chem. Phys. **98**, 5648 (1993)
29. Y. Zhao, D.G. Truhlar, Theor. Chem. Acc. **120**, 215 (2008)
30. R.G. Parr, W. Yang, *Density-Functional Theory of Atoms and Molecules* (Oxford University Press, 1989)
31. J.P. Perdew, K. Burke, M. Ernzerhof, Phys. Rev. Lett. **77**, 3865 (1996)
32. G. Scalmani, M.J. Frisch, Chem. Theory Comput. **8**, 4494 (2012)
33. J. Heyd, G.E. Scuseria, M. Ernzerhof, J. Chem. Phys. **118**, 8207 (2003)
34. F. Aryasetiawan, O. Gunnarsson, Rep. Prog. Phys. **61**, 237 (1998)

Chapter 11
Integration of Density Functional Theory with Machine Learning: An Overview

Muhammad Raziq Rahimi Kooh, Yuan-Fong Chou Chau, Chee Ming Lim, and Roshan Thotagamuge

Abstract This chapter explores machine learning (ML) integration into density functional theory (DFT) and addresses challenges related to accuracy, computational cost and scalability. It provides a comprehensive overview of different types of ML models and algorithms, followed by discussions on ML-driven improvements to DFT components such as the exchange–correlation functionals, basis sets, self-consistent field and integration grid using various ML techniques. The applications of different ML models, including supervised, unsupervised and semi-supervised learning that replace direct DFT calculations, are also discussed. Case studies highlight the significant impact of ML-DFT integration across various domains, from electronic properties and catalytic reaction predictions to screening materials with the desired properties. Despite these advancements, challenges remain in the ability of some ML models to generalise with new data and scale effectively, highlighting the opportunities for further development and innovation.

Keywords Machine learning · Density functional theory · Exchange–correlation functionals · Surrogate model · High-throughput screening · Computational materials discovery

Chee Ming Lim was Deceased on 12/01/2026.

M. R. R. Kooh (✉) · Y.-F. Chou Chau · C. M. Lim
Centre for Advanced Material and Energy Sciences, Universiti Brunei Darussalam, Universiti Brunei Darussalam, Gadong, Brunei Darussalam
e-mail: raziq.kooh@ubd.edu.bn

R. Thotagamuge (✉)
Department of Optometry, Faculty of Allied Health Sciences, University of Sri Jayewardenepura, Nugegoda, Sri Lanka
e-mail: roshan@sjp.ac.lk

R. Thotagamuge (ed.), *Density Functional Theory*,
https://doi.org/10.1007/978-981-95-9745-1_11

11.1 Introduction

Density Functional Theory (DFT) is a computational tool commonly used to study the electronic properties of molecules and materials. Instead of solving the complex Schrödinger equation directly, DFT simplifies the problem by focusing on the electronic density, which makes the calculations much faster and more practical [1]. This approach allows DFT to predict electronic structures, reaction mechanisms and material properties at a relatively low computational cost [1]. As a result, it has contributed to the advancement of various domains ranging from drug design to renewable energy technologies [2, 3].

DFT plays a vital role in understanding atomic reaction mechanisms and facilitating virtual screening for new materials, such as battery components, to improve performance and reduce development costs. However, to achieve accurate and reliable results, several key components of DFT play a crucial role in determining the accuracy of the calculations.

The exchange–correlation (XC) functional is one of the key components of DFT that affects the accuracy of DFT calculations [4]. The XC functional is significant in describing the many-body interactions between electrons. Within the Kohn–Sham (KS) framework, the total energy of a system is expressed as a sum of non-interacting electron energies, with the electron–electron interactions encapsulated in the XC functional. However, the exact form of exchange–correlation energy density (ε_{xc}) is unknown and, therefore, must be approximated, making the choice of XC functional critical for the accuracy of DFT predictions [5]. In this framework, the KS eigenvalues and wavefunctions are key outputs that depend on the form of the XC functional. The KS eigenvalues correspond to the energy levels of the electrons in the system, and the KS wavefunctions describe their spatial distribution. These quantities are crucial for determining the electron density, which governs the system's total energy and derived properties. The highest occupied KS eigenvalue (HOMO) corresponds to the negative ionisation potential, but errors in the XC functional approximation can lead to deviations from experimental values. More accurate functionals improve KS eigenvalues and wavefunctions, reducing errors in band gap predictions, ionisation energies and charge distribution [6].

Apart from XC functionals, another key component of DFT is the self-consistent field (SCF) procedures. SCF refers to the iterative method to solve the KS equations while achieving the desired accuracy and managing computational costs. The goal is to find a self-consistent electron density that minimises the system's total energy. It starts with an initial guess for the electron density, which is then used to solve the KS wavefunctions and eigenvalues. Convergence is checked by comparing the new density with the previous one. Convergence criteria are used to determine when the SCF procedure should stop, typically based on thresholds for changes in total energy and electron density between iterations. The procedure is terminated when the density and energy converge to a stable solution [7]. The number of iterations depends on the system's complexity. A larger system will incur higher computational cost, as more iterations may be required to reach self-consistency.

Another component is the numerical integration grid, which is essential for calculating the energy functional, particularly for electron–electron interactions and XC functionals. The grid is constructed by combining atomic grids into a molecular grid with optimised angular and radial components. The grid resolution affects both accuracy and computational cost. Finer grids improve accuracy but increase computational demands, while coarser grids reduce cost but may lead to errors [8–10].

The basis set is also an important DFT component, consisting of functions representing the electronic wavefunctions. A good choice of basis set is crucial for ensuring accurate wavefunction representation, which directly impacts the calculation's accuracy. The larger and more flexible the basis set, the more accurate the results, but it also increases computational demand.

The rise of open data repositories and the growing focus on data-driven approaches drive the rapid growth and adoption of machine learning (ML). ML is a subset of artificial intelligence that enables computers to learn from data and make predictions or decisions without being explicitly programmed. In recent years, many studies have focused on integrating ML with DFT to address some of DFT's limitations. The goal is to leverage large datasets to train ML models, enabling more efficient and accurate predictions of material properties and reaction outcomes.

The primary objective of applying ML to DFT is to improve the accuracy and efficiency of DFT calculations. While DFT is generally more efficient than other quantum–mechanical methods, such as energy-direct tree search and coupled-cluster methods, it remains computationally expensive, particularly for large systems or complex calculations [11]. Integrating ML with DFT offers significant potential to improve accuracy and reduce computational requirements. Supervised and unsupervised ML techniques can be leveraged to improve these [12, 13].

A key challenge in DFT is selecting the appropriate XC functional, which significantly impacts the accuracy of the results. Traditional functionals are based on theoretical approximations, and while they perform well for many systems, they may fail for specific types of materials or reactions, particularly when complexity increases. ML technique presents a promising approach to improving these functionals' accuracy by identifying patterns and correlations that enable the development of more accurate, data-driven functionals. Apart from direct enhancement of the XC functional, ML can also be incorporated into other components of DFT, such as the use of ML to produce a high-quality initial guess of electron density that minimises the number of iterations of the SCF cycle [7], development of an adaptive basis set using ML technique to replace traditional basis set [14], and an ML-improved integration grid [8].

Also, using ML to train surrogate models with large datasets that approximate DFT results offers a much faster alternative to direct DFT calculations. These MLs create predictive models by identifying patterns and correlations that yield more accurate predictions without running DFT calculations, which would otherwise require extensive computational resources. Another exciting ML application in conjunction with DFT is deep learning, where the molecules are represented as graphs and linked to accurate DFT data. This approach enables high-throughput screening, which evaluates many materials or compounds to identify promising candidates for a particular application, thereby accelerating the discovery of new materials for use in batteries, solar cells or catalysts.

This work explores integrating ML techniques with DFT to address key challenges in computational chemistry. Section 11.2 provides an overview of ML techniques, introducing different learning types, algorithms and data structures. Section 11.3 explores the use of ML to improve multiple aspects of DFT. Section 11.4 discusses the coupling of ML and DFT to train surrogate models that rapidly identify materials for an intended application without costly direct DFT calculations.

11.2 Machine Learning Basics

11.2.1 Supervised, Unsupervised, Semi-supervised and Reinforcement Learning

ML models are generally categorised into four main types: supervised, unsupervised, semi-supervised and reinforcement learning [15]. Each ML learning category addresses different challenges and learning scenarios. They involve distinct methods, algorithms and applications that enable models to learn from data and make predictions. In the following section, each learning technique is discussed regarding its relevance and applicability when integrated with DFT.

11.2.1.1 Supervised Learning

Supervised learning involves training the model on a labelled dataset, where the input and outcome variables are known. It includes various algorithms for classification and regression. Classification refers to predicting a discrete category or label, while regression predicts continuous values. Some of the common supervised learning algorithms are regression methods (e.g. linear regression, logistic regression (LgR)), neural networks (e.g. artificial neural networks), decision trees (DT), random forests (RF), k-nearest neighbours (kNN) and support vector machines (SVM).

Supervised learning algorithms have been applied to improve DFT calculations by enhancing efficiency, reducing computational costs and training ML models to predict material properties that would otherwise require expensive calculations.

Linear Regression

Linear regression predicts a continuous value by fitting a straight line to the data. It assumes a linear relationship, in which one variable (the dependent variable) is predicted from the value of the other (the independent variable). The line is commonly represented by the equation $y = mx + c$ where y is the dependent variable, x is the

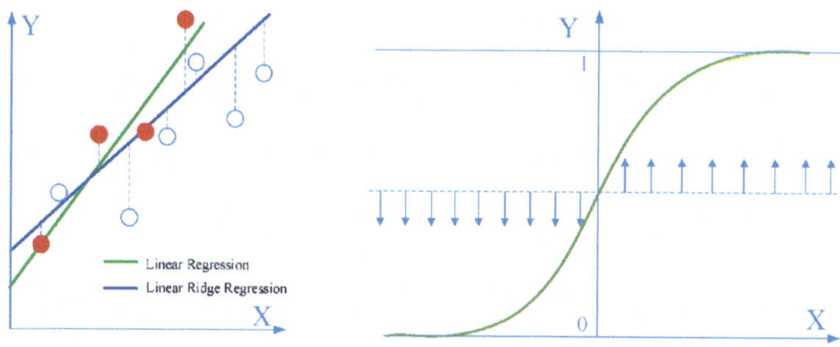

(a) LRR model. (b) LgR model.

Fig. 11.1 Graphical representation showing the concepts of **a** linear regression that fits a straight line that best represents the input, linear ridge regression where a penalty (regularisation) is added to prevent overfitting by shrinking the coefficients of the model, and **b** logistic regression where the model fits a sigmoid curve to the data, producing a probability value between 0 and 1. If the value exceeds a certain threshold, it leads to a prediction output of class 1; otherwise, class 0, making it suitable for binary classification tasks [19]. Used under Creative Commons CC BY 4.0

independent variable, m is the gradient or slope of the line, and c is the y-intercept (Fig. 11.1a). The goal is to find the best-fitting line that minimises the difference between the observed and predicted values, typically using the least squares method [16].

Linear regression is too simple to handle tasks of higher complexity, and in most cases, it can suffer from overfitting. Overfitting occurs when a model learns the details and noise in training data so well that it performs poorly on unseen data, making it less able to adapt to new situations [17]. Regularisation techniques can help address this issue, particularly when the model has multiple features or when the data are highly multicollinear.

Ridge regression is an extension of linear regression that includes a penalty term to control the magnitude of the model coefficients (Fig. 11.1a). This regularisation is added to the loss function, which helps prevent overfitting by shrinking the coefficient [16]. Kernel ridge regression (KRR) extends ridge regression by combining it with kernel tricks.

Logistic Regression

LgR is a statistical method for binary classification that predicts the probability of one of two possible outcomes. LgR estimates probabilities using the sigmoid function, which models the relationship between input features and the output by applying a linear combination of the features (Fig. 11.1b). This combination then passes through the sigmoid function to produce a probability value between 0 and 1. If the probability exceeds a certain threshold, the prediction is class 1; otherwise, it is class 0. Although named 'regression', it is primarily used for classification tasks [16]. LgR itself is not very useful in DFT; however, its variants, multiple logistic

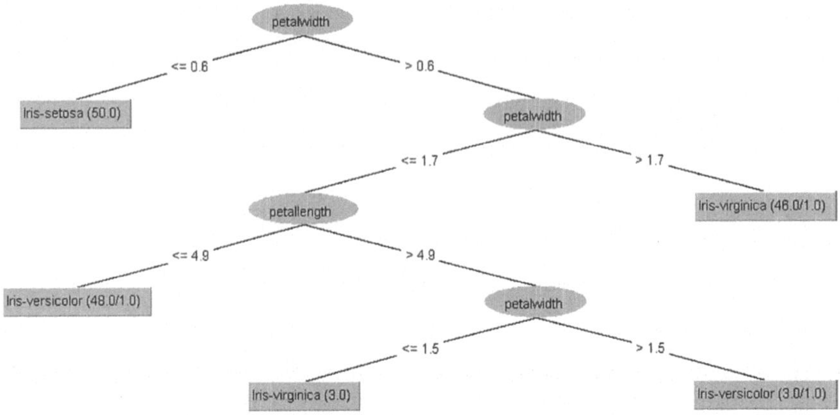

Fig. 11.2 Visualisation of the decision tree model trained on the iris dataset using Weka. The dataset consists of measurements for three iris species (Setosa, Versicolor and Virginica), with each sample featuring four attributes: sepal length, sepal width, petal length and petal width. The values at the splits represent the threshold values used to partition the data at each decision node [22]

regression (MLgR), which extend to the prediction of multiple binary outcomes, have proven useful in DFT applications [18].

Decision Tree

A DT creates a tree-like structure for decision-making based on features and can be used for classification and regression tasks [20]. It works by recursively asking a series of questions about the data's features, with each question leading to a branch based on the answers (Fig. 11.2). These questions are structured as nodes, and paths taken through the tree's hierarchy ultimately lead to a classification label or a value [21]. DT is easy to interpret, as it breaks down decisions into a series of simple, understandable rules. However, they are prone to overfitting if too complex, which can be mitigated through pruning or limiting tree depth. DT can provide more robust and accurate predictions when combined in ensembles, such as random forests or boosting methods [21].

Random Forests

RF is an ensemble method that combines multiple decision trees (a forest), using bootstrap aggregating (bagging) and either majority voting (for classification) or averaging (for regression) to aggregate their predictions. This ensemble approach often leads to better accuracy when compared to a single decision tree. Each tree in the forest is trained on a random subset of the data, and a random subset of features is used to split each node. This randomness reduces overfitting and ensures that individual trees in the forest are less correlated, thereby improving generalisation. As the number of trees increases, the generalisation error tends to converge to a limit. The more trees, the more stable the model becomes [23, 24].

kNN

kNN is a simple, non-parametric algorithm for classification and regression. The term non-parametric means the algorithm makes predictions based on the distances between data points rather than fitting a parameter-based model (Fig. 11.3). A commonly used distance metric is the Euclidean distance. The algorithm's prediction for a data point is based on the majority class (for classification) or the average value (for regression) of its k-nearest neighbours in the feature space [25]. As a lazy learner, kNN defers learning until prediction time. Instead of building a model during training, it stores the entire training dataset. It only computes distances to the stored data when a new data point needs to be classified or predicted. This approach contrasts with eager learning algorithms, which perform computations beforehand. kNN is also known as an instance-based learning algorithm or memory-based learning. However, kNN has a high calculation cost when using a large dataset [26].

Support Vector Machine

Vapnik laid the theoretical groundwork for support vector machines (SVMs) in the 1960s, but the algorithms were developed and popularised in the 1990s [28, 29]. The SVM algorithm finds the best boundary or optimal hyperplane between different categories while maximising the margin between them (Fig. 11.4a). The lines adjacent to the optimal hyperplane are known as the margins. A larger margin results in better generalisation of unseen data. Since their introduction in the 1990s, SVMs have been used for weather prediction, power estimation, stock prediction, optical character recognition, audio processing and medical diagnosis [30].

This method is adapted for regression, known as support vector regression (SVR), by introducing an ε-insensitive loss function that allows the model to predict continuous values while maintaining robustness to small variations in the data. The ε refers to the tolerance zone around the regression line (hyperplane) where errors are not penalised. The purpose of ε is to allow some error or deviation between the predicted and true values without affecting the model's performance [31]. If data points lie within the ε margin, they are correctly predicted [32] (Fig. 11.4b).

While both the margins in SVM and SVR are related to tolerances, they serve different purposes in classification versus regression problems.

The kernel was later integrated into the algorithm to enable non-linear classification and regression. The kernel is a weighting function for a sum or an integral. It transforms the non-linear input space into a high-dimensional feature space, where the solution to the problem can be represented as a linear classification or regression problem (Fig. 11.5) [32]. Some popular kernels used in SVMs are the radial basis

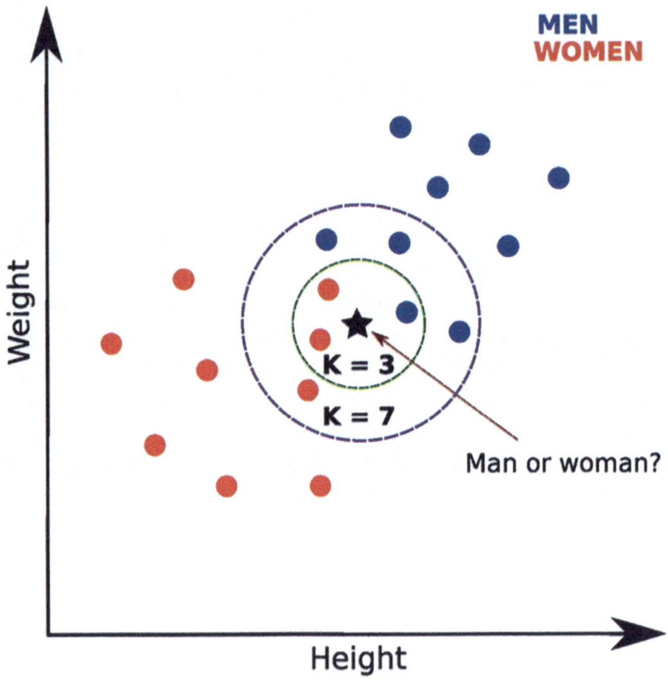

Fig. 11.3 A kNN representation illustrates individuals' classification as male or female based on weight and height [27]. This diagram shows the kNN classification process with two different values of K. In the green circle ($K = 3$), where the classification is based on the 3 nearest neighbours: 2 red circles (female) and 1 blue circle (male), leading to a classification of female. In the purple circle ($K = 7$), where the classification is based on the 7 nearest neighbours, 3 red circles and 4 blue circles, a classification of a male is obtained. This example demonstrates how kNN assigns class labels based on the majority of the nearest neighbours in the feature space. Used under Creative Commons CC BY 4.0

function kernel, the universal Pearson VII function-based kernel and the polynomial kernel [32].

Neural Networks Algorithms

Neural networks (NN) are a class of algorithms inspired by the function of biological neurons. Currently, many variants of NN are applied to both classification and regression tasks.

An artificial neural network (ANN) is a widely used NN algorithm. An ANN is a network of interconnected neurons (nodes), typically organised in an input layer, one or more hidden layers and an output layer (Fig. 11.6). In a feedforward NN, information flows in one direction (unidirectional), from the input layer through hidden layers to the output layer, without looping [34]. During training, backpropagation

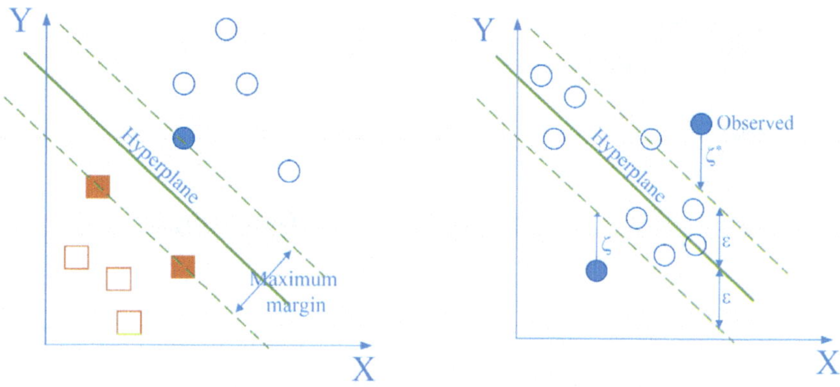

(**a**) SVM. (**b**) SVR.

Fig. 11.4 Graphical representation of **a** support vector machine where the goal is to find a hyperplane that best separates data points of different classes with the maximum margin and **b** support vector regression where the goal is to find a function that best fits the data, ensuring that most data points fall within a specified margin (ε), while minimising prediction error (ζ) for points outside the margin [19]. Used under Creative Commons CC BY 4.0

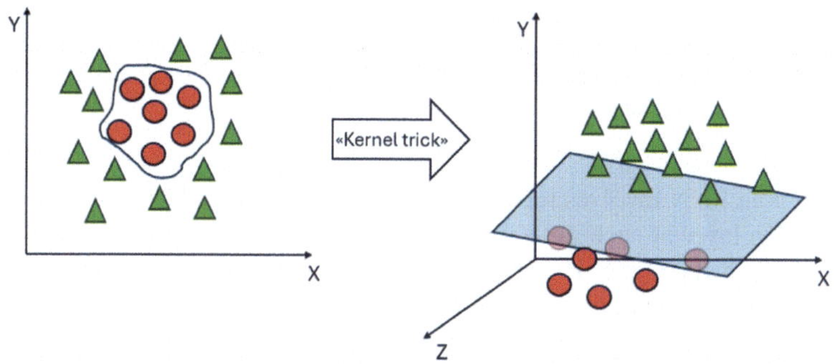

Fig. 11.5 Illustration of the kernel trick, which transforms data into a higher-dimensional space to enable the support vector machine to find an optimal hyperplane for classification. This technique separates non-linearly separable data by mapping it to a higher-dimensional feature space [33]. Used under Creative Commons CC BY 4.0

calculates the error between the predicted output and the actual data and propagates it backwards through the network [34]. The algorithm then adjusts the network's weights using this error signal, typically via gradient descent, allowing the network to improve its accuracy iteratively [34]. ANN has been reported to be coupled with DFT in many research studies [35].

A recurrent neural network (RNN) is a specific type of NN designed to handle sequential and time-dependent data. Unlike a regular feedforward neural network, which only passes information from one layer to the next, RNNs have recurrent connections (a feedback loop) that allow information to cycle within the network.

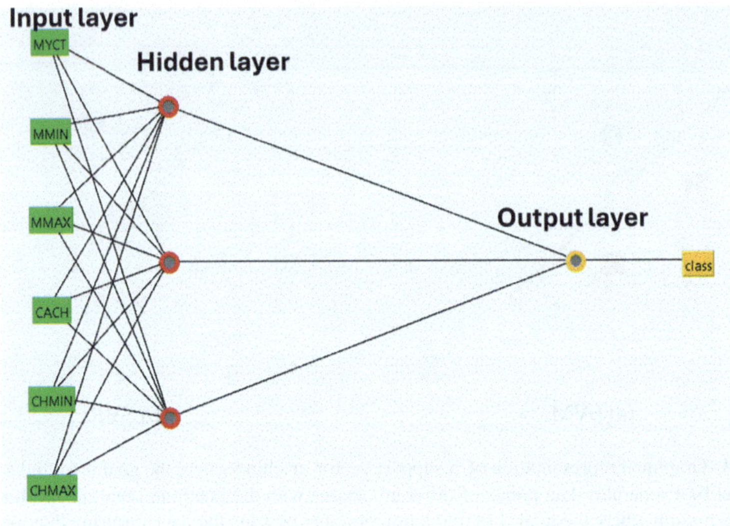

Fig. 11.6 Visualisation of the artificial neural network (ANN) model trained on CPU performance data using Weka software, with performance evaluated across different CPU architectures based on various attributes [22]. The diagram depicts the architecture of the trained ANN, which consists of an input layer representing the CPU attributes such as the machine cycle time (MYCT), minimum main memory (MMIN), maximum main memory (MMAX), cache memory size (CACH), minimum cache size (CHMIN) and maximum cache size (CHMAX), a hidden layer that processes these inputs and an output layer that predicts CPU performance

The core of an RNN involves recurrent computation of the hidden state (the output of the hidden layer at each time step), which integrates the current input with the previous hidden state [36]. This design enables the RNN to remember and use past inputs for future predictions. It makes them particularly well-suited for tasks where the order of data points and their context are critical, such as time-series forecasting [37].

Graph neural networks (GNNs) are specialised neural networks for graph-structured data. A GNN consists of nodes (or vertex) and edges (or links), forming a graph together (Fig. 11.7). A node represents an attribute or an entity, and an edge represents the relationship between attributes and specifies direction, where the edge can be undirected or directed [38]. The prediction can be at the node, edge or graph levels, making it valuable in fields such as chemistry, drug discovery and materials science to predict molecular properties with the molecule as a graph, nodes as the atoms and each edge representing a bond between atoms, capturing the molecule's structure [38]. GNNs are often called representational learning methods and can be supervised or unsupervised. The core mechanism of GNNs consists of two phases: (1) message passing, where each node updates its representation by receiving information from its neighbouring nodes using a learned function, and (2) a readout phase, where the final node states are aggregated to generate molecular property predictions

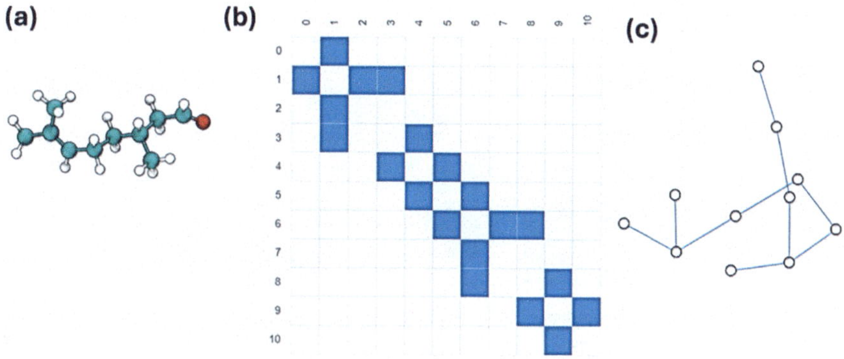

Fig. 11.7 Diagrams showing the **a** 3D representation of citronella, illustrating its spatial configuration, **b** adjacency matrix of the bonds in the molecule, which represents the connectivity between atoms based on the bonds in the molecule and **c** graph representation of the citronella molecule, where the atoms are nodes, and bonds are edges connecting these nodes, providing a network-like representation of the molecular structure [38]. Used under Creative Commons CC BY 2.0

[39]. A reported variant of GNN used to improve DFT is a Message Passing Neural Network (MPNN) [39].

Convolutional neural networks (CNNs) are another specialised type of NN, originally designed for image classification and processing grid-like data (Fig. 11.8) [40]. CNN uses filter optimisation to learn features. It is composed of layers, including convolution layers, pooling layers and fully connected layers [41]. Using layers in a CNN enables it to handle complex, non-linear relationships between the input and output without requiring a predefined model or function [42]. A convolutional layer contains kernels (or filters) with adjustable parameters. These kernels slide across the image, computing a dot product between the pixel values and the kernel's weights. The outcome is a convolved image [42]. CNN learns spatial hierarchies of features through a backpropagation algorithm [43]. CNNs have been reported to predict DFT energies with high accuracy [42].

A convolutional graph neural network (CGNN) extends the concept of convolution from grid-like data (e.g. images) to graph-structured data (Fig. 11.9). It applies polynomial filters to the neighbourhoods of nodes in the graph, aggregating information from neighbouring nodes to update node representations. This method allows the network to process graphs with irregular structures, making it suitable for node classification and link prediction in graph-based data [45].

Gradient Boosting Machines

Gradient Boosting Machines (GBMs) are machine learning algorithms that build an ensemble of models one at a time. Each new model is trained to correct the errors made by the previous one. The process relies on gradient descent, where each model is adjusted to minimise the loss function, improving the overall prediction (Fig. 11.10).

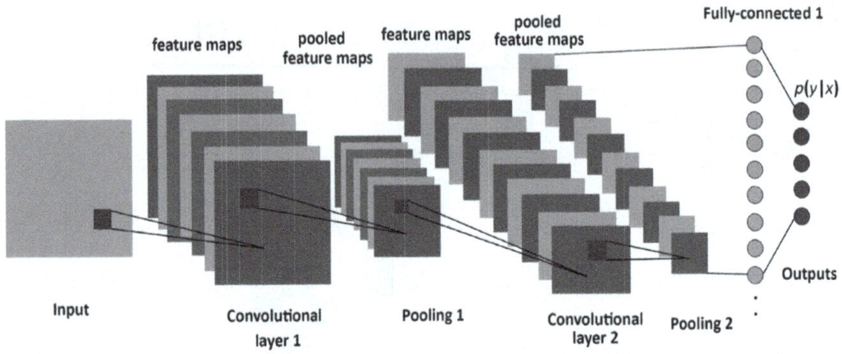

Fig. 11.8 Schematic diagram of a Convolutional Neural Network (CNN) [44]. This diagram illustrates the architecture of a CNN, which consists of convolutional layers that extract features from the input, pooling layers that reduce spatial dimensions, and the resulting pooled feature maps. The learned features are then passed through fully connected layers for classification or regression tasks. Used under Creative Commons CC BY 4.0

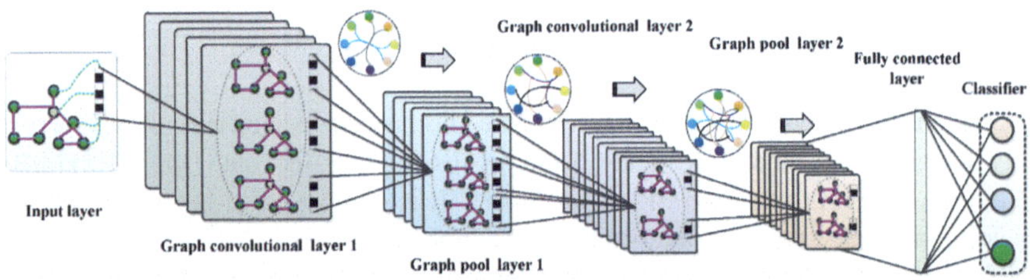

Fig. 11.9 The diagram illustrates the architecture of a Graph Convolutional Neural Network (GCNN) that operates on a graph [46]. The graph convolutional layer processes the node features by aggregating information from neighbouring nodes. The graph pooling layer reduces the complexity of the graph data, and the resulting pooled feature maps are then passed through a fully connected layer for classification or regression. Used under Creative Commons CC BY 4.0

The loss function can be chosen based on the specific problem, making GBMs versatile and customisable. This flexibility, combined with their strong performance in various applications, has made GBMs popular for accurate predictive modelling for classification and regression tasks [47]. Variants such as XGBoost (Extreme Gradient Boosting) and LightGBM have been reported to be used in DFT-related work [48].

XGBoost is a high-performance implementation of the gradient-boosting framework designed for efficiency and scalability. It incorporates advanced optimisation techniques, such as a sparsity-aware algorithm and a weighted quantile sketch, enabling it to handle sparse data and perform approximate tree learning. XGBoost is widely used in machine learning tasks, particularly for classification and regression, thanks to its ability to deliver high-quality predictive models. It also supports distributed computing, making it suitable for large-scale datasets [50].

LightGBM is a gradient-boosting framework that leverages decision trees for efficient, fast computation and better memory efficiency, especially with large datasets.

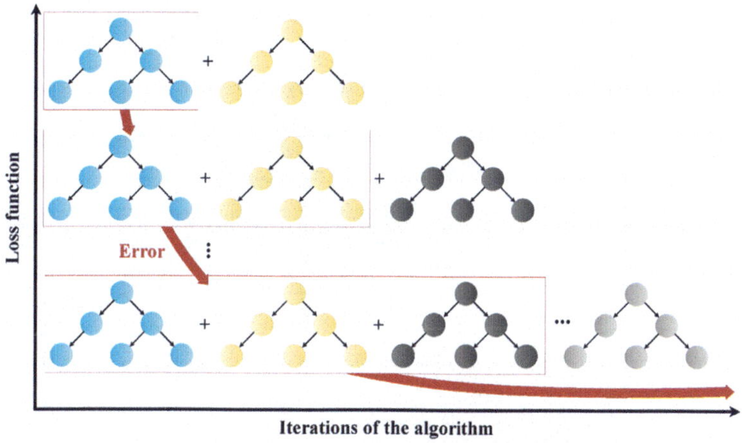

Fig. 11.10 Illustration of the Gradient Boosting technique, where multiple weak learners are iteratively trained to correct the errors of previous models. The loss function decreases with each iteration, reflecting the model's accuracy over time as the combined output of the trees forms a more accurate predictive model [49]. Used under Creative Commons CC BY 4.0

LightGBM, designed to overcome the limitations of XGBoost, has been reported to outperform most GBMs and to exhibit high scalability [51, 52]. LightGBM also supports classification and regression tasks.

Naive Bayes

Naive Bayes is a probabilistic classifier based on Bayes' theorem, assuming independence between features. It calculates the probability of each class given the input features and selects the class with the highest probability.

11.2.1.2 Unsupervised Learning

Unsupervised learning identifies patterns from unlabelled data. Examples of unsupervised learning include clustering and dimensionality reduction.

K-means Clustering

K-means clustering is one of the oldest clustering algorithms used to group similar data points into distinct clusters. It works by assigning each data point to one of the K clusters based on how close it is to the cluster's centre, known as the centroid. The algorithm starts by randomly selecting K centroids, then iteratively adjusts them

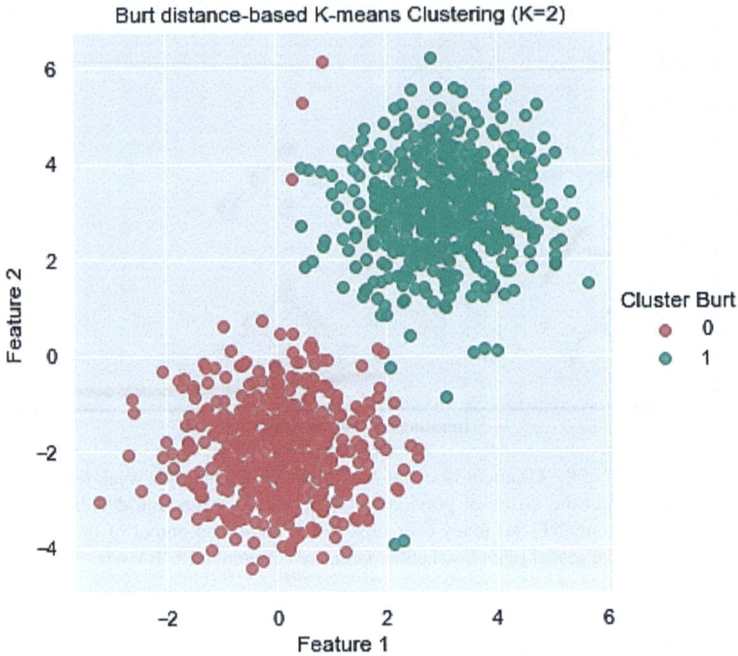

Fig. 11.11 Illustration of *K*-means clustering using Burt distance, where data points are grouped into clusters based on their similarity as measured by the Burt distance metric [53]. The red and blue circles represent two distinct clusters, with points of the same colour belonging to the same group. Used under Creative Commons CC BY 4.0

by minimising the distance between each data point and its assigned centroid. This process continues until the clusters converge (Fig. 11.11) [53].

Hierarchical Clustering

Hierarchical clustering groups data into clusters based on shared characteristics. The grouping is based on distance similarity and produces a list of different levels of clustering. This list is represented as a dendrogram (Fig. 11.12) [54].

Principal Component Analysis

Principal component analysis (PCA) is a dimensionality reduction technique for simplifying complex datasets by transforming them into a new set of uncorrelated variables, called principal components, that capture the most significant variance in the data. PCA is very useful for preprocessing data before applying other algorithms and was reported in many DFT-related publications [42, 55].

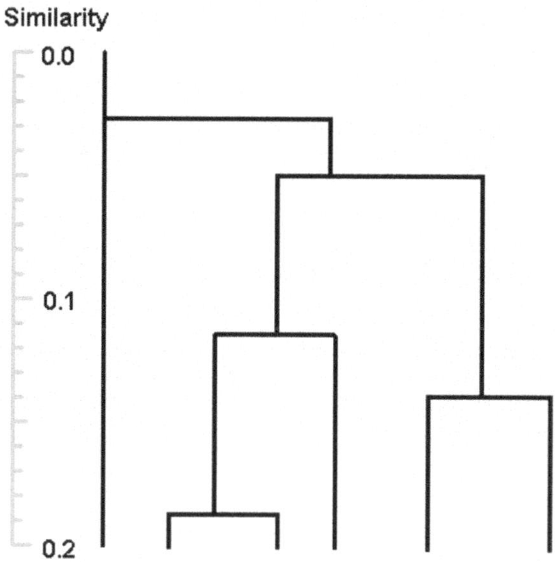

Fig. 11.12 A dendrogram represents hierarchical clustering, where data points are progressively merged into clusters based on similarity. The tree-like structure illustrates the order and distance at which clusters are combined, with each branch's height indicating the similarity level between clusters. The closer the clusters are merged (near 0), the more similar they are, and the larger the distance (towards 0.2), the more different the clusters are from each other [54]. Used under Creative Commons CC BY 4.0

11.2.1.3 Semi-supervised Learning

Semi-supervised learning is a hybrid of supervised and unsupervised learning that combines labelled and unlabelled data. The model is trained on a small labelled dataset and leveraged on a larger pool of unlabelled data to improve learning accuracy (Fig. 11.13). By utilising unlabelled data, semi-supervised learning can address the small-sample-size problem, which occurs when the number of labelled training samples is insufficient relative to the feature dimension, thereby improving classification performance [56]. One key aspect of semi-supervised learning is its ability to automatically select reliable training data from the unlabelled pool. Unsupervised clustering and self-training are commonly employed, where a classifier iteratively selects confident samples to add to the training set. Co-training, which involves multiple classifiers that share new training samples, is another popular approach. One open-source tool for DFT applications that employs semi-supervised learning is SemiH [57].

11.2.1.4 Reinforcement Learning

Reinforcement learning trains agents to make decisions by interacting with an environment to maximise cumulative reward. The agent learns through trial and error and optimises its behaviour based on feedback in the form of rewards or penalties resulting from its actions (Fig. 11.14). Several works have explored the use of reinforcement learning in developing ML surrogate models [58, 59].

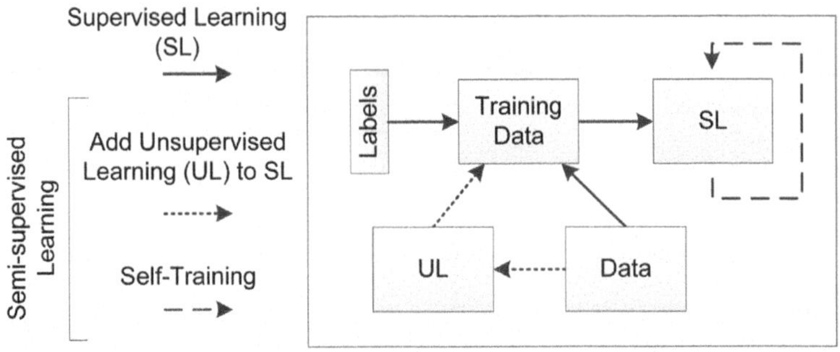

Fig. 11.13 A representation of semi-supervised learning illustrates the process by which a model learns from labelled and unlabelled data. A small labelled dataset guides the learning process, while a larger unlabelled dataset helps improve the model's accuracy and generalisation [56]. Used under Creative Commons CC BY 4.0

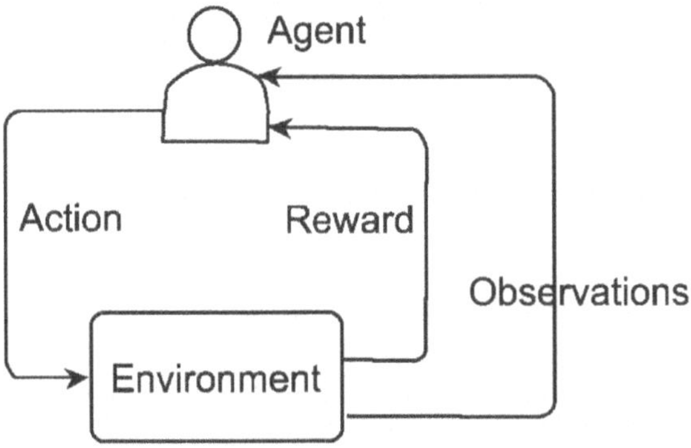

Fig. 11.14 Illustration of agent-environment interaction in reinforcement learning, where an agent interacts with its environment by taking actions based on the current state. The environment provides feedback through rewards or penalties, which the agent uses to adjust its behaviour and decisions over time [60]. Used under Creative Commons CC BY 4.0

11.2.2 Data Pre-processing

Data preprocessing is an essential step before training an ML model, and it can significantly affect the quality of the results.

11.2.2.1 Missing Data

Datasets can have partial or incomplete data entries, leading to missing values in one or more variables. One of the most common methods for handling missing data is to delete them using deletion methods [61], such as list-wise deletion (removing rows with missing values) and pair-wise deletion (excluding only the missing data points). Another technique is imputation, where missing values are reconstructed by assigning them the mean or median, by regressive imputation that predicts values based on other features, or by a more advanced method such as kNN imputation, which uses nearby values to estimate missing data [62].

11.2.2.2 Normalisation of Data

Normalisation of data is important because data for each variable can be at different scales or orders of magnitude. Normalisation reduces data complexity and ensures that all dataset features have a similar distribution range.

The most common scaler is the MinMaxScaler, where data is transformed to values between 0 and 1 using Eq. 11.1 [63]:

$$x_N = \frac{(x_i - x_{\min})}{(x_{\max} - x_{\min})} \tag{11.1}$$

where x_N is the normalised value, x_i is the original data, while $x_{\min}$ and $x_{\max}$ are the lowest and highest values in the column of the original data, respectively.

StandardScaler is another scaler where the data is transformed to have a mean (μ) of 0 and a standard deviation (σ) of 1 using Eq. 11.2 [63]:

$$x_N = \frac{(x_i - \mu)}{\sigma} \tag{11.2}$$

RobustScaler is often used for datasets with many outliers and scales the data based on the median and interquartile range (IQR) rather than the minimum and maximum values. The equation is as shown in Eq. 11.3 [63].

$$x_N = \frac{(x_i - \text{Median}(x))}{IQR} \tag{11.3}$$

11.2.2.3 Data Augmentation

Data augmentation refers to adding more instances to an existing dataset to mitigate the risk of overfitting, particularly when the dataset is small. It also helps increase dataset diversity and allows the model's ability to generalise better [64]. In the context of a DFT dataset, additional training data can be generated by varying the atomic configurations of molecules by rotating, shifting or substituting atoms or perturbing the DFT-calculated properties, such as bond lengths, angles and electronic properties to simulate variations in a real-world system [65].

11.2.2.4 Featurisation

Featurisation is useful in coupling DFT with ML by transforming raw molecular data, such as molecular structures or atomic configurations, into a machine-readable numerical or categorical format [52]. This step ensures that the data is in a form that the models can process and learn from effectively. RDKit, Open Babel and the Chemistry Development Kit (CDK) are popular cheminformatics toolkits for molecular data preprocessing and featurisation.

However, it also increases data complexity by adding high-dimensional features representing various molecular characteristics. Recursive feature elimination, a function in the scikit-learn ML library, can further curate the molecular fingerprinting data by removing irrelevant and redundant features [52].

11.2.3 Training-Test Split and Cross-Validation Techniques

In ML, it is essential to partition the input data into training and testing datasets. Training data teaches models to make predictions by identifying patterns, relationships and trends. The test data is used to evaluate the model's performance. Partitioning ensures that models can generalise effectively and avoid overfitting specific data points.

11.2.3.1 Percentage Split

A percentage split is a simple approach in which the dataset is randomly divided into training and test subsets. A typical ratio of 70% for training and 30% for testing works well when large datasets are available and is equally common in the 80:20 split. However, there is a risk of bias in the split, where the training or test set may not capture the full diversity of the data. This can lead to poor performance, especially when the data are not well randomised. A common error often observed in the work of inexperienced researchers is applying a percentage split on a small dataset without ensuring randomisation, resulting in unrepresentative splits and unreliable models with poor performance.

11.2.3.2 Cross-Validation

Cross-validation (CV) is a technique that can mitigate the risk of a percentage split. It serves dual purposes: ensuring the model is trained on all input data and evaluating its performance to provide a more comprehensive assessment of its generalisability.

Some common variants of CV are K-fold CV, stratified K-fold CV and Leave-One-Out CV (LOO-CV).

In K-fold CV, the dataset is divided into K equal-sized folds. Each iteration uses one fold for testing while the model is trained on the other K-1 folds. This cycle repeats for each fold, ensuring that every data point is included in both the training and testing phases. The splits are random, and there is no guarantee that the class distribution in each fold matches the overall distribution across the entire dataset (Fig. 11.15). Another close variant of CV is stratified K-fold CV, which preserves the class distribution in each fold and is particularly important for imbalanced datasets [17].

LOO-CV is a special case of K-fold CV where K equals the number of data points (n). In LOO-CV, one data point is left out for testing while the model is trained on the remaining n-1 data points. This process is repeated for each data point in the dataset. LOO-CV is attractive because it provides the maximum amount of data for

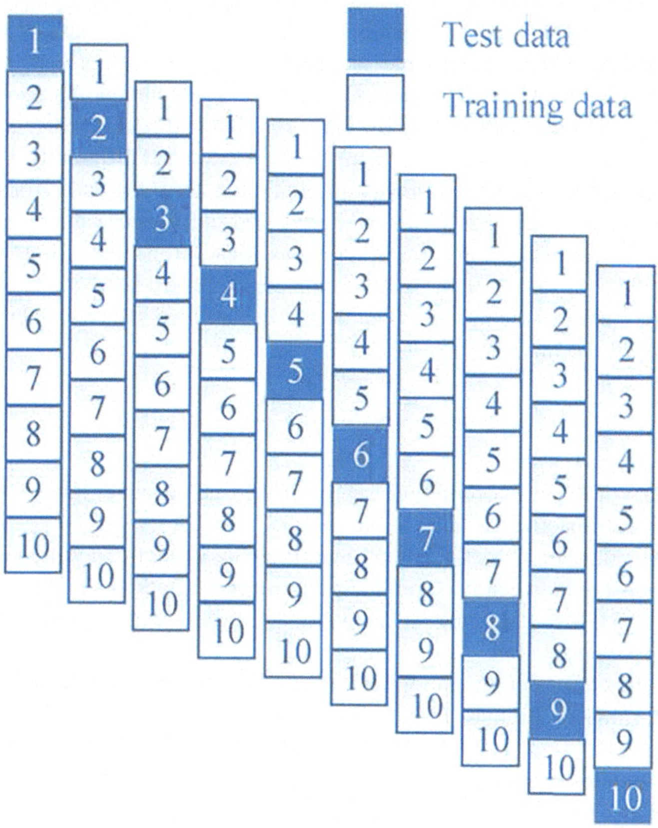

Fig. 11.15 A graphical representation of a typical k-fold cross-validation, where, in this example, $k = 10$. The process divides the dataset into 10 equal folds, where each fold is used for validation while the remaining are used to train the model. The process is repeated 10 times, ensuring each data point is used for training and validation, and the final model performance is averaged across all iterations [19]. Used under Creative Commons CC BY 4.0

training and increases the likelihood that the ML model is accurate. LOO-CV is also deterministic, with no random sampling and no need for repetition. LOO-CV can be computationally expensive and may not be practical for large datasets; however, it offers an opportunity to squeeze the most out of a small dataset [17].

A common mistake made by inexperienced researchers is the redundant use of both a percentage split and a CV. For example, applying a 70:30 training-to-test split in conjunction with a tenfold CV creates unnecessary duplication of effort. This is because CV has already divided the data into training and test folds, rendering the initial percentage split redundant.

11.2.4 Evaluation of Models

The evaluation of ML models depends on whether the task is classification (e.g. predicting whether a bond is cleavable) or regression (e.g. predicting energy values or electronic properties). Classification models assess the likelihood of an event occurring and often produce discrete outputs, while regression models predict continuous numerical values. Classification and regression are evaluated differently. For classification tasks, evaluation metrics focus on prediction accuracy and error rates; for regression tasks, the difference between the predicted and original values is measured. While there are many common evaluators for classification and regression, only those reported in the case studies in the later section are discussed here.

Accuracy measures the proportion of correctly classified instances relative to the total number of instances.

$$\text{Accuracy} = \frac{\text{TP} + \text{TN}}{\text{TP} + \text{TN} + \text{FP} + \text{FN}} \tag{11.4}$$

where TP, TN, FP and FN represent true positive, true negative, false positive and false negative, respectively.

Pearson's correlation coefficient (r) measures the strength and direction of the linear relationship between two variables: the original values (x_i) and the predicted values (y_i) [17]. The value ranged between -1 and 1, with a value closer to 0 indicating poor correlation.

The coefficient of determination (R^2) is another common metric for evaluating model performance. It ranges from 0 to 1, with values closer to 1 indicating better performance.

Mean absolute error (MAE) is the average of the absolute differences between the original and predicted data. It measures how far off the prediction is on average and is less sensitive to outliers when compared to mean squared error (MSE) or root mean squared error (RMSE). The smaller the MAE, the better the prediction [66].

The average error ratio (AER) is the average error in model predictions, used to measure the relative deviation of predicted values from the original values.

MSE is the average squared difference between a model's original and predicted values. It measures the variance of the residuals and gives more weight to large errors, making it more sensitive to outliers than MAE. A lower MSE indicates better performance. While MSE is useful, it is harder to interpret directly due to its squared units.

RMSE is a square root of MSE, which measures the standard deviation of residuals. Similar to MAE and MSE, a smaller value indicates a smaller error. RMSE is usually preferred as it has the same unit as the dependent variable [17, 66].

The equations of the performance evaluators are as follows [17, 66]:

$$\text{Pearson's } r = \frac{\left(\sum_{i=1}^{n}(x_i - \bar{x})(y_i - \bar{y})\right)}{\sqrt{\sum_{i=1}^{n}(x_i - \bar{x})^2 \sum_{i=1}^{n}(y_i - \bar{y})^2}} \tag{11.5}$$

$$R^2 = 1 - \frac{\sum_{i=1}^{n}(y_i - x_i)^2}{\sum_{i=1}^{n}(x_i - \bar{x})^2} \tag{11.6}$$

$$\text{MAE} = \frac{1}{n}\sum_{i}^{n}|y_i - x_i| \tag{11.7}$$

$$\text{AER} = \frac{1}{n}\sum_{i}^{n}\frac{|y_i - x_i|}{x_i} \tag{11.8}$$

$$\text{MSE} = \frac{1}{n}\sum_{i}^{n}(y_i - x_i)^2 \tag{11.9}$$

$$\text{RMSE} = \sqrt{\frac{1}{n}\sum_{i}^{n}(y_i - x_i)^2} \tag{11.10}$$

where x_i and y_i represent the value of the original and predicted data, respectively, $\bar{x}$ and $\bar{y}$ are the means of the original and predicted values, and n is the number of data points.

11.3 Improvement of DFT Using ML Techniques

Recent advancements in DFT software and tools have increasingly incorporated ML techniques to address the longstanding challenges in accuracy, efficiency and computational cost, paving the way for more precise and scalable electronic structure calculations.

In ORCA 5.0, the developers redesigned the integration grid to overcome the limitations of the original linear-scaling grid, which no longer met modern computational chemistry standards [8]. The grid serves as a 3D mesh spanning the space around a molecule, with calculations performed at specific points. Increasing the number of grid points improves accuracy but increases computational costs. To strike a balance, the new grids (*Defgrid1*, *Defgrid2* and *Defgrid3*) were optimised using ML to achieve higher accuracy and efficiency. These grids are also self-adapting, meaning they automatically allocate more computational effort to challenging regions of the molecules, such as areas with steep or diffuse functions, directing computational effort where it is most needed. This adaptability ensures more reliable and precise calculations compared to older versions of ORCA [8].

GradDFT is a JAX-based software library (a Python library for high-performance numerical computing) designed for rapid prototyping and experimentation with differentiable functionals. It aims to improve the accuracy of the DFT, particularly for transition metals and strongly correlated systems, by leveraging neural functionals [13]. The accuracy of DFT calculations depends heavily on the approximation used for the XC functional—a core of DFT that approximates electron exchange and correlation interactions. Traditional functionals use hand-crafted expressions for the exchange–correlation energy density (ε_{xc}), while GradDFT replaced this with a data-driven model using a neural network (supervised learning) to parameterise the XC functional, allowing it to learn ε_{xc} from a large dataset and capture more complex electron-correlation effects more accurately. Neural networks compute the coefficients in the XC functional as functions of electron density and its derivatives, incorporating higher-order dependencies without being constrained by fixed analytical forms. This allows GradDFT to go beyond traditional functionals, such as generalised gradient approximations (GGAs) and meta-GGAs, and to learn a more flexible representation of XC energy terms [13]. Unlike conventional functionals, which derive the exchange–correlation potential (Vxc) from predefined forms, GradDFT parameterises Vxc directly using a neural network, ensuring self-consistency and improving electron density predictions. The framework is fully differentiable, allowing gradient-based optimisation of functionals and enabling direct learning of Kohn–Sham orbitals without relying on precomputed electron densities. By leveraging automatic differentiation in JAX's and optimising functionals with various loss functions, GradDFT achieves high accuracy while remaining computationally scalable through multi-GPU and TPU acceleration [67]. Its self-consistent training approach refines density predictions iteratively within the SCF cycle, yielding more accurate and adaptive functionals. As a result, GradDFT enhances energy calculations and electron density predictions, offering a more accurate and efficient alternative to traditional DFT methods.

GradDFT demonstrated how neural networks can improve the accuracy of XC functionals by replacing traditional approximations with a data-driven approach. Beyond this methodology, other researchers have explored hybrid strategies that integrate supervised and unsupervised learning techniques. Kovács, for instance, combined neural networks with k-means clustering and PCA to improve functional performance and dataset balance [12]. Kovács first applied unsupervised learning techniques to analyse and pre-process the dataset. K-means clustering groups materials with similar electronic properties, such as those with comparable band gap trends and cohesive-energy variations. This clustering revealed inherent trends in how different materials respond to variations in the XC functional, allowing for a more informed selection of training data. By identifying clusters of materials that occupy similar regions in the descriptor space, Kovács was able to rebalance datasets to minimise biases and improve the representativeness of the training data for functional development. PCA was further used to reduce the dataset's dimensionality, ensuring that the most relevant descriptors, such as the Laplacian of the electron density and the kinetic energy density, were retained while removing redundant or weakly contributing variables. This dimensionality reduction ensured that the subsequent supervised learning process focused on the most informative features, improving the efficiency and accuracy of the functional training. For the supervised learning component, Kovács trained 25 meta-GGA functionals using a dataset of 440 materials for band gap predictions and 44 materials for cohesive-energy and lattice-parameter optimisations. The functionals were assigned different weightings to examine trade-offs among the accuracy of lattice parameters, cohesive energies and bandgaps. The training process used a flexible meta-GGA functional form, and the loss function was a weighted combination of errors in lattice parameters, cohesive energies and bandgaps. For instance, the trained functionals outperformed existing meta-GGA functionals such as SCAN and PBE in predicting bandgaps, with some achieving mean absolute relative errors as low as 31.75%, compared to 35.22% for SCAN and 47.03% for PBE. However, the trained functionals showed that improving one property often came at the expense of another, a limitation that persists even with ML-assisted approaches. For example, functionals optimised for cohesive-energy tended to perform poorly in predicting lattice parameters and vice versa. This underscores the need for more sophisticated weighting strategies or hybrid approaches to achieve balanced performance across multiple properties. By integrating supervised and unsupervised ML techniques, Kovács developed a more systematic and data-driven approach to XC functional optimisation. This approach not only improved DFT accuracy for a broader range of materials but also provided valuable insights into the limitations of traditional functionals in capturing multi-property correlations [12, 68].

The accuracy of DFT calculations depends significantly on the choice of basis sets, which approximate the wavefunctions of electrons in a system. Schütt and VandeVondele developed an adaptive basis set using ML techniques that can improve the efficiency and accuracy of DFT calculations [14]. Traditional atom-centred basis sets, such as Gaussian-type orbitals (GTO) and plane-wave approaches, are static and isotropic, leading to inefficiencies in large-scale simulations. Their proposed adaptive basis set is based on projected atomic orbitals (PAOs) that dynamically adjust based on the local chemical environment. A Gaussian Process Regression (GPR) model was trained on explicitly optimised PAO basis sets, learning how to refine the localised basis functions for various atomic configurations. This data-driven optimisation eliminates manual basis-set tuning while preserving the accuracy of larger, more computationally intensive basis sets. The model was trained using DFT-calculated energies and forces, and its performance was validated through molecular dynamics simulations of liquid water. Minimal adaptive basis sets reduce computational cost by a factor of 200 while maintaining accuracy comparable to that of larger basis sets [14]. The ML-based basis adaptation eliminates the need for iterative basis optimisation, making large-scale DFT simulations more feasible.

In another work on developing an adaptive basis set, Khan et al. employ KRR to predict optimal scaling factors for the radial exponents of contracted GTOs, allowing the basis functions to dynamically adjust to the local chemical environment [69]. The training data consists of scaling factors optimised through variational minimisation of Hartree–Fock energies for small molecules from the QM9 database covering carbon, hydrogen, oxygen, nitrogen and fluorine. The ML model is trained separately for each element. The ML-enhanced basis sets were evaluated by comparing total energies, atomisation energies, bandgaps and dipole moments against larger, more accurate basis sets such as cc-pVQZ. The adaptive basis sets reduced atomisation energy error and improved accuracy in 99% of the test cases. These improvements were achieved with negligible computational overhead, making the ML-optimised basis sets an efficient alternative for large-scale calculations.

Further integrating ML into DFT calculations, Hazra et al. reported using neural networks to predict the one-particle density matrix (DM), thereby significantly improving computational efficiency [7]. Electronic structure calculations typically require an iterative process known as an SCF cycle, which refines the electron density until it reaches a stable ground state. The number of iterations depends on the initial guess for the DM or electron density. A high-quality initial guess using ML reduces the number of iterations needed for convergence. Hazra et al. trained a neural network on 9000 configurations, using 800 for validation and 1000 for testing [7]. The resulting ML model requires only the molecule's Cartesian coordinates as input and predicts the elements of the DM. For simple molecules such as H_2O and S_2O, the ML-generated DM could converge in just one iteration, whereas traditional methods require 3–5 iterations [7]. This approach significantly reduces the computational cost of electronic structure calculations.

Collins and Raghavachari introduced an ML framework that significantly enhances the accuracy of DFT for thermochemical calculations. By leveraging the systematic error-cancellation properties of the Connectivity-Based Hierarchy fragmentation scheme, they developed molecular descriptors that encode the local chemical environment of atoms. These descriptors were used in a supervised learning model, specifically a feedforward neural network, to predict and correct the errors in

DFT calculations. The resulting ML-corrected DFT models achieve chemical accuracy (errors within 0.5 kcal/mol) for a wide range of organic molecules, making DFT a more reliable tool for thermochemical predictions without requiring computationally expensive high-level quantum-chemical methods. This approach bridges the gap between the high accuracy of high-level quantum chemistry and DFT's computational efficiency, offering a practical solution for accurate thermochemical studies.

Similarly to Collins and Raghavachari, Kalita et al. also used a supervised learning model, a feedforward neural network, to approximate XC functionals by incorporating density errors in the loss function [55]. Known physical constraints were integrated into the neural network, enabling accurate approximations with limited training data. KRR was used to approximate kinetic energy functionals in early ML-DFT efforts. At the same time, low-dimensional structures in electron density data were captured using kernel PCA and non-linear gradient denoising [55]. All these lead to a key development, the Kohn–Sham regulariser (KSR), which enhances self-consistency by treating the XC functional as a differentiable component in neural networks. This eliminates the need for expensive self-consistent cycles.

From another direction, Stoltz and Borunda proposed a post-processing correction to improve the accuracy of DFT-calculated bandgaps and ionisation potentials for organic semiconducting polymers, introducing a simple linear regression-based method [70]. The authors benchmark six XC functionals (three LDA and three GGA) and find that PBE and PW91 offer the best predictive performance. A simple linear correction function is derived from a training set of 22 polymers, reducing systematic DFT errors and achieving average percentage errors (APE) below 3% for bandgap and 4% for ionisation potential. The corrected values demonstrate that a simple machine-learning-assisted calibration can significantly enhance the reliability of DFT predictions.

By integrating ML, modern DFT software and tools are now faster, more accurate and better equipped to tackle complex systems, marking a significant leap forward in computational chemistry.

11.4 Integration of ML into DFT as a Computation Shortcut

The previous section described how ML can enhance DFT's accuracy and computational efficiency by improving XC functionals and related methods. However, ML also serves as a powerful computational shortcut, enhancing DFT workflows without replacing them. This section explores how material screening and discovery can be accelerated by using ML models to analyse DFT-derived data and predict key properties, significantly reducing computational costs. The case studies of ML-DFT integration discussed in this section are briefly summarised in Table 11.1.

Günay et al. used MLgR to predict O_2 adsorption stability and strength on gold clusters. Their dataset consisted of 97 configurations with 7 descriptors (4

Table 11.1 Overview of ML-DFT coupling configurations

Nature of work	ML models	Number of features/descriptors	Total number of instances	Number of test data	Performance evaluator	Output	References
Ligand prediction for a catalyst	ANN	15	21	1	LOO-CV, MSE, R^2	Predict catalyst performance	Fan et al. [35]
Ligand design	ANN	11	22	1	LOO-CV, R^2	Predict the selectivity of the ligand	Zhang et al. [71]
Total energy and surface properties of nanoalloy	ANN, specifically NNP	592	20,000	2000	RMSE, MAE	Predict total energy and surface properties of Au/Cu NPs	Artrith and Kolpak [72]
O_2 on gold clusters	MLgR	7	92	-	fourfold CV, Chi-square test, Walk test	Classify adsorption as stable or unstable and weak or strong	Günay et al. [18]
Fluorescent dyes	RF	35	1000 to 10,000	NS	Pearson R RMSE	Excitation energies and oscillator strengths	Kang et al. [73]
Nitric oxide reduction reaction to ammonia	XGBoost	13	27	NS	RMSE, R^2	Overpotential value	Liu et al. [2]
TiNi-based shape memory alloys	XGBoost	7	207	NS	LOO-CV, Pearson R, R^2	thermal hysteresis	Tian et al. [74]
Polymer bandgap	LGB-Stack	600	4209	20%	R^2, RMSE	Predict bandgap	Goh et al. [52]
Polymer bandgap	SVR	1093	284	20%	R^2, RMSE, LOO-CV	Predict bandgap	Xu et al. [75]

(continued)

Table 11.1 (continued)

Nature of work	ML models	Number of features/descriptors	Total number of instances	Number of test data	Performance evaluator	Output	References
Molecular property prediction	GNN, specifically MPNN	23	130,462 (10,000 randomly chosen for training)	10,000 randomly chosen for testing	Average error ratio	Predict quantum properties	Gilmer et al. [39]
Total energy of atomistic systems	CNN	Image-based descriptor	269,016	30% of dataset	MAE	Predict the total energy of 2D hexagonal lattices	Ryczko et al. [42]
Photovoltaic material discovery	Hierarchical Clustering	32	540	NA	NS	Identified compound for photovoltaic use	Saha et al. [76]
DFT Hamiltonian	SemiH	NS	10–30 labelled, 60–300 unlabelled	40–50	MSE	Predict DFT Hamiltonian	Cho et al. [57]
Crystal design	Offline reinforcement learning	Graph-based features	8832	2486	Accuracy (%)	Stable crystalline material with a desired bandgap	Govindarajan et al. [59]

NS means 'not specified', and NA means 'not applicable'.

user-defined descriptors: cluster size, charge, coordination number and presence of unpaired electrons and 3 electronic properties: bandgap, ionisation potential and binding energy). The training data were derived from DFT calculations, allowing the MLgR model to pre-screen and identify the key descriptors that most significantly influence adsorption sites before performing computationally expensive DFT simulations [18].

While a single DT may not improve DFT, the RF algorithm shows considerable promise. Using the RF algorithm, Kang et al. predicted the excitation energies and oscillator strengths of fluorophores used in bioimaging [73]. The model is trained using molecular descriptors represented by extended connectivity fingerprints (a type of molecular fingerprint) and an additional 36 molecular descriptors, such as molecular weight, number of rings, fraction of SP3 carbon, etc. The ECFP provides a compact, detailed representation of the molecular structure, while the additional molecular descriptors offer broader, more general characteristics of the molecules. The model's strong performance, with an RMSE of 0.46 and a Pearson R of 0.87, underscores its ability to generalise well with new data and accurately predict key properties that enhance the design of bioimaging dyes.

ANN is a popular ML model widely used for training DFT-related models. Fan et al. constructed an ANN model to design asymmetric PNP ligands for ethylene oligomerisation catalysts [35]. In their model, the ANN was tested across various configurations for its accuracy in predicting the catalyst's performance. Their work used a dataset compiled from their previous research, totalling 21 instances, with steric and electronic descriptors obtained from DFT calculations. For a small dataset, LOO-CV was computationally affordable and used, and the model obtained a very high R^2 of 0.98. The model was further optimised by adjusting the number of neurons in the hidden layer and achieved an R^2 of 0.99. Another work by the same research group uses an ANN to predict the selectivity of silicon-bridged diphosphine ligands in the ethylene tri-/tetramerisation reaction, with steric and electronic descriptors extracted from DFT calculations, and achieves high prediction accuracy ($R^2 = 0.99$) [71]. While a high R^2 may suggest a strong predictive model on a given dataset, it may also indicate the possibility of overfitting, especially with a limited number of instances. While LOO-CV may minimise the risk of overfitting, it does not eliminate it entirely. Additional performance metrics, such as MAE and RMSE, can provide a more holistic view of model performance, and sensitivity analyses, such as SHAP (Shapley Additive Explanations), can offer deeper insights into feature importance and prediction reliability. The small datasets in these studies could benefit from data augmentation to enhance diversity and improve the model's generalisation.

Artrith and Kolpak combined DFT with a variant of ANN to overcome the computational challenges of simulating large gold-copper (Au/Cu) nanoalloys. The ANN used is the Parrinello–Behler Neural Network Potential (NNP), which is designed for atomistic simulation in materials science and computational chemistry. The model was trained on 20,000 DFT-derived structures, with 2000 test data. The resulting model achieved an RMSE of 8.2 meV/atom and an MAE of 6.0 meV/atom. They concluded that NNP can replicate DFT accuracy while reducing computational cost by several orders of magnitude [72]. This highlighted the potential of ML in replacing expensive DFT calculations while maintaining high accuracy, particularly when trained on a large dataset.

Solving the KS equation is a primary computational bottleneck in DFT, and traditional solvers scale poorly with the system size. Chandrasekaran et al. introduced an ML-based surrogate model that efficiently predicts electron density and local density of states (LDOS) from atomic configurations without explicitly solving the KS equations [77]. By bypassing direct KS equation solving, ML enables linear scaling with system size, making large-scale electronic structure calculations feasible. A fully connected neural network (for charge density) and a recurrent neural network (RNN) for LDOS were trained on millions of DFT-generated data points obtained from molecular dynamics simulations of aluminium and polyethene. Training used a dataset of eight snapshots, validation used one snapshot, and testing used a completely unseen configuration, ensuring robust model generalisation within the same material system; however, it does not explicitly test the model's ability to predict electron densities for unseen materials with different chemical compositions. The ML approach was validated using R^2 and RMSE, and the model achieved near-perfect agreement with DFT, with R^2 values of 0.999997 for polyethene and 0.999955 for aluminium. However, such high R^2 values raise concerns about overfitting and the model's ability to generalise beyond the training data.

XGBoost has also gained significant popularity in the field. Liu et al. demonstrated a fast ML-based screening approach for catalyst discovery, specifically for designing efficient single-atom alloy catalysts to reduce nitric oxide to ammonia [2]. They generated DFT calculations for various catalysts and trained an XGBoost model using 13 descriptors and only 27 instances. The model achieved an R^2 of 0.95 and RMSE of 0.33, demonstrating XGBoost's ability to accelerate the search for optimal catalysts without relying on exhaustive DFT computations [2]. However, the limited dataset size limits the diversity of training samples and potentially affects extrapolation to novel catalyst compositions. Expanding the dataset through data augmentation could improve the model's robustness.

Using a similar ML algorithm, Tian et al. developed a data-driven approach for designing shape memory alloys (SMAs) with optimised thermal properties, focusing on reducing thermal hysteresis (T_{hys})—a critical property for elasto-caloric cooling applications [74]. The model was trained on a dataset of 207 SMAs with a key physical property, λ_2 eigenvalue, as a critical feature to enhance prediction accuracy [74]. By incorporating LOO-CV, the model achieved an R^2 score of 0.87. Testing 500 randomly generated TiNiCu components reveals that increasing the Cu content to 15% yields the lowest T_{hys}. While the study successfully identified promising compositions, further experimental validation is essential to confirm ML-driven predictions. All this work demonstrates that coupling XGBoost with DFT accelerates material discovery by screening promising candidates before full DFT simulations, significantly reducing computational costs.

Goh et al. demonstrated in their study that LGB-Stack, a two-level stacked generalisation model using LightGBM, could predict polymer bandgap values [52]. There are several key findings from their work. The feature redundancy resulting from featurisation can be effectively trimmed using recursive feature elimination without sacrificing accuracy. The LGB-Stack model also achieved an R^2 score of 0.92 and an RMSE of 0.41 at an 80:20 train: test split, indicating high accuracy and outperforming the benchmark QSPR model.

Also, on polymer bandgap prediction, Xu et al. used the SVR algorithm to train a dataset containing 1093 descriptors and 284 instances [75]. The SVR model with LOO-CV yields an R^2 of 0.824 with a Gaussian kernel. The SVR model was further optimised by conducting a tree of Parzen Estimator algorithm, and the final model achieved an R^2 of 0.925. Despite having a high R^2, the model's predicted bandgap deviated significantly from the DFT-calculated values. The authors highlighted the lack of data on high-bandgap polymers as the model's most significant limitation, affecting its ability to generalise well.

Zhang et al. conducted an interesting study that predicted photocatalytic cleavage of C–C bonds in lignin [78]. They first classify whether bond cleavage is likely to occur (binary classification: 'cleavable' vs 'non-cleavable') using five ML models: kNN, naïve Bayes, SVM, LgR and RF, where all the models exhibited good accuracy at 0.90, 0.91, 0.97, 0.96 and 0.99, respectively. Instead of running expensive DFT calculations on all bonds, the classifiers help pre-screen which bonds are likely to cleave under given photocatalytic conditions, allowing them to focus on the most relevant cases. After classification, RF was used to train a surrogate model to predict bond dissociation energy associated with C–C bond cleavage.

While RF, ANN, GBM and SVR have tremendous accuracy, one drawback is the increased computational cost for large datasets, which may be required to achieve high accuracy. Saha et al. reported that hierarchical clustering, an unsupervised learning technique, can help materials scientists with small datasets overcome this constraint [76]. Their study applied an agglomerative hierarchical clustering algorithm to analyse 540 halide double perovskites. The model identified 10 clusters and enabled the researchers to select 10 compounds with potential photovoltaic applications for further DFT analysis, revealing that 8 of the compounds meet the desired bandgap range, highlighting the model's high accuracy. Similarly, Mai et al. use K-means clustering to analyse DFT data to identify clusters of photocatalysts with desirable properties [79].

Graph neural network techniques offer many opportunities in computational chemistry. Gilmer et al. introduced Message Passing Neural Networks (MPNNs) as a deep learning approach for predicting quantum-chemical properties directly from molecular graphs [39]. Their method leverages graph neural networks to model atomic interactions, replacing traditional feature engineering with learned message-passing mechanisms. By training on the QM9 dataset, which contains DFT-calculated molecular properties, MPNNs effectively approximate these quantum–mechanical results at a fraction of the computational cost [2]. While MPNNs do not replace DFT as a fundamental theory, this approach is a fast surrogate model that enables high-throughput property screening before complete quantum–mechanical calculations are performed. The synergy lies in its ability to learn from DFT-derived data, enabling large-scale screening of molecules that would otherwise be computationally prohibitive with traditional quantum chemistry methods.

Ryczko et al. accelerate total-energy calculations for atomistic systems by using a CNN to approximate DFT-calculated energies [42]. In this approach, the CNN learns from DFT-calculated energies and predicts them at a fraction of the cost, avoiding expensive direct DFT calculations. Trained on DFT-based datasets of graphene, hBN and their heterostructures, the CNN model can estimate DFT energies with an MAE of 0.198 meV/atom, where this small error margin indicates that the model can capture the physics of the atomic systems very effectively. The model also shows strong generalisation capabilities, accurately predicting energies in new molecular dynamics simulations.

Cho et al. presented an innovative semi-supervised approach to develop SemiH, an open-source surrogate model based on a DFT Hamiltonian neural network that replaces direct DFT calculations [57]. The training dataset included labelled atomic configurations paired with their corresponding DFT Hamiltonians, as well as unlabelled atomic configurations without DFT data. This approach demonstrates enhanced generalisation, especially for more complex datasets, such as larger atomic systems and higher temperatures. Key findings include significant reductions in Hamiltonian and band-structure errors, with SemiH achieving up to 58% lower Hamiltonian errors than traditional methods. The study concludes that SemiH is a reliable tool for accurate DFT Hamiltonian predictions, offering reduced computational costs and enabling broader applications in materials science and semiconductor device design.

Several studies have used reinforcement learning to couple ML with DFT data. Meldgaard et al. predict the atomistic structure of a reconstructed crystalline surface using a deep neural network as a reinforcement learning agent, interacting with an environment that includes a quantum–mechanical potential-energy evaluator based on DFT [58]. The agent treats the 3D atomistic structure as a series of stacked 2D images and learns to place atoms and select atomic sites. The agent requires between 1000 and 10,000 single-point DFT evaluations to autonomously determine the optimal surface reconstructions of anatase and rutile tin oxide [58].

Similarly, Govindarajan et al. developed a reinforcement learning framework for designing stable crystalline materials with a targeted bandgap, evaluated via DFT [59]. The agent learns to fill atoms into a predefined crystal skeleton to optimise for low formation energy and a target band gap. The reward function incorporates DFT-computed properties, and the model is trained using offline reinforcement learning on a dataset of 8,832 non-metallic crystals from the Materials Project. The generated materials are validated using DFT, and the results show that the conditioned policies can design crystals with the desired properties. While this technique demonstrated accuracy, relying on DFT for both the reward function and the final evaluation makes this approach computationally expensive.

Integrating ML with DFT has significantly improved computational efficiency for materials discovery, catalyst screening and property prediction. However, different ML models have distinct strengths and limitations when used as surrogate models. At the same time, some models struggle with data generalisation, and others, such as reinforcement learning, face challenges in scaling to complex systems, highlighting the need for further optimisation and dataset expansion.

11.5 Conclusion

Integrating ML with DFT revolutionises computational chemistry by enhancing accuracy, efficiency and predictive capabilities. ML improves the DFT components through direct functional and basis-set enhancements, improved integration grids and reduced self-consistent-field iterations, significantly reducing the computational cost of DFT calculations while maintaining accuracy. Using ML to train surrogate models from DFT-derived data enables the prediction of desired material properties, avoiding costly direct DFT calculations. Moreover, training ML models to screen materials enabled selecting a few that are most likely to have the desired properties for DFT calculations. Some ML models face generalisation challenges, while others, particularly reinforcement learning models, remain too costly to scale up. These challenges present opportunities for further development and innovation.

References

1. R. Pederson, B. Kalita, K. Burke, Machine learning and density functional theory. Nature Reviews Physics **4**(6), 357–358 (2022)
2. J. Liu et al., *Understanding the activity origin of Pd-anchored single-atom alloy catalysts for NO-to-NH3 conversion by DFT studies and machine learning,* (Chinese Chemical Letters, 2024), p. 110656
3. M.M. Taouti et al., DFT and machine learning integration to predict efficiency of modified metal-free dyes in DSSCs. J. Mol. Graph. Model. **136**, 108975 (2025)
4. Q. He et al., Density Funct. Theory Batter. Mater. **2**(4), 264–279 (2019)
5. M.K. Caucci et al., Performance of exchange-correlation approximations to density functional theory for rare-earth oxides. Comput. Mater. Sci. **253**, 113837 (2025)
6. R. Stowasser, R. Hoffmann, What do the Kohn– Sham orbitals and eigenvalues mean? J. Am. Chem. Soc. **121**(14), 3414–3420 (1999)
7. S. Hazra, U. Patil, S. Sanvito, Predicting the one-particle density matrix with machine learning. J. Chem. Theory Comput. **20**(11), 4569–4578 (2024)
8. F. Neese, Software update: The ORCA program system—Version 5.0. Wiley Interdiscip. Rev.: Comput. Mol. Sci. **12**(5), e1606 (2022)
9. Gaussian. *Integral—Grid options*. 2022 28 February 2022 19 Mar 2025, https://gaussian.com/integral/?tabid=1#Integral_keyword__UltraFine_Option.
10. *7.3. Details on the numerical integration grids*. ORCA 6.0 Manual 2024 19 Mar 2025, https://www.faccts.de/docs/orca/6.0/manual/.
11. J.R. Reimers, *Computational methods for large systems: electronic structure approaches for biotechnology and nanotechnology*, (Wiley, 2011)
12. P. Kovács, *Machine learning application for DFT exchange functionals*, (Technische Universität Wien, 2021)
13. P.A.M. Casares et al., GradDFT. A software library for machine learning enhanced density functional theory. J. Chem. Phys. **160**(6) (2024)
14. O. Schütt, J. VandeVondele, Machine learning adaptive basis sets for efficient large scale density functional theory simulation. J. Chem. Theory Comput. **14**(8), 4168–4175 (2018)
15. I.H. Sarker, Machine learning: algorithms, real-world applications and research directions. SN Comput. Sci. **2**(3), 160 (2021)
16. J. Starmer, *The StatQuest Illustrated Guide to Machine Learning: Triple Bam*, (StatQuest Publications, 2022)

17. I.H. Witten et al., Chapter 5—Credibility: evaluating what's been learned, in *Data Mining*, ed by I.H. Witten, et al., 4th edn. (Morgan Kaufmann, 2017), pp. 161–203
18. M.E. Günay, T. Davran-Candan, R. Yıldırım, Analysis of O2 adsorption stability and strength over gold clusters using DFT and logistic regression. J. Cluster Sci. **23**(2), 221–235 (2012)
19. J.-S. Chou, D.-N. Truong, C.-F. Tsai, Solving regression problems with intelligent machine learner for engineering informatics. Mathematics. **9** (2021). https://doi.org/10.3390/math9060686
20. M. Xu et al., Decision tree regression for soft classification of remote sensing data. Remote Sens. Environ. **97**(3), 322–336 (2005)
21. C. Kingsford, S.L. Salzberg, What are decision trees? Nat. Biotechnol. **26**(9), 1011–1013 (2008)
22. M. Hall et al., The WEKA data mining software: an update. ACM SIGKDD Explor. Newsl **11**(1), 10–18 (2009)
23. L. Breiman, Random forests. Mach. Learn. **45**(1), 5–32 (2001)
24. M.R.R. Kooh et al., Machine learning approaches to predict adsorption capacity of Azolla pinnata in the removal of methylene blue. J. Taiwan Inst. Chem. Eng. **132**, 104134 (2022)
25. O. Kramer, K-Nearest neighbors, in *Dimensionality Reduction with Unsupervised Nearest Neighbors*, (Springer Berlin Heidelberg, Berlin, Heidelberg, 2013), pp. 13–23
26. M. Punjabi, G.L. Prajapati, *Lazy learner and PCA: An evolutionary approach*, (IEEE, 2017)
27. D. Lopez-Bernal et al., Education 4.0: teaching the basics of KNN, LDA and simple perceptron algorithms for binary classification problems. Futur. Internet. **13** (2021). https://doi.org/10.3390/fi13080193
28. A.J. Smola, B. Schölkopf, A tutorial on support vector regression. Statistics computing **14**, 199–222 (2004)
29. V.N. Vapnik, Pattern recognition using generalised portrait method. Autom. Remote. Control. **24**(6), 774–780 (1963)
30. M. Awad, R. Khanna, *Efficient learning machines: theories, concepts, and applications for engineers and system designers*, (Springer Nature, 2015)
31. S.K. Shevade et al., Improvements to the SMO algorithm for SVM regression. IEEE Trans. Neural Netw. **11**(5), 1188–1193 (2000)
32. B. Üstün, W.J. Melssen, L.M.C. Buydens, Facilitating the application of support vector regression by using a universal Pearson VII function based kernel. Chemom. Intell. Lab. Syst. **81**(1), 29–40 (2006)
33. R. Guido et al., An overview on the advancements of support vector machine models in healthcare applications: a review. Information **15** (2024). https://doi.org/10.3390/info15040235
34. J.A.L. Marques et al., Artificial neural network-based approaches for computer-aided disease diagnosis and treatment, in *Cognitive and Soft Computing Techniques for the Analysis of Healthcare Data*, ed. by A.K. Bhoi, V.H.C. de Albuquerque, Srinivasu, P. N., G. Marques, (Elsevier, 2022), pp. 79–99
35. H. Fan et al., Computer-assisted design of asymmetric PNP ligands for ethylene tri-/tetramerization: a combined DFT and artificial neural network approach. J. Catal. **418**, 121–129 (2023)
36. I.D. Mienye, T.G. Swart, G. Obaido, Recurrent neural networks: a comprehensive review of architectures, variants, and applications. Information **15** (2024). https://doi.org/10.3390/info15090517
37. H. Salehinejad et al., *Recent advances in recurrent neural networks*. arXiv:1801.01078 (2017)
38. B. Sanchez-Lengeling et al., A gentle introduction to graph neural networks. Distill **6**(9), e33 (2021)
39. J. Gilmer, et al., *Neural message passing for quantum chemistry*. in *International conference on machine learning*, (PMLR, 2017)
40. A. Khan et al., A survey of the recent architectures of deep convolutional neural networks. Artif. Intell. Rev. **53**, 5455–5516 (2020)
41. S.H.S. Basha et al., Impact of fully connected layers on performance of convolutional neural networks for image classification. Neurocomputing **378**, 112–119 (2020)

42. K. Ryczko et al., Convolutional neural networks for atomistic systems. Comput. Mater. Sci. **149**, 134–142 (2018)
43. R. Yamashita et al., Convolutional neural networks: an overview and application in radiology. Insights Imaging **9**(4), 611–629 (2018)
44. S. Albelwi, A. Mahmood, *A framework for designing the architectures of deep convolutional neural networks*. Entropy **19** (2017). https://doi.org/10.3390/e19060242
45. A. Daigavane, B. Ravindran, G. Aggarwal, Understanding convolutions on graphs. Distill **6**(9), e32 (2021)
46. X. Zhao et al., Intelligent health assessment of aviation bearing based on deep transfer graph convolutional networks under large speed fluctuations. Sensors **23** (2023). https://doi.org/10.3390/s23094379
47. A. Natekin, A. Knoll, Gradient boosting machines, a tutorial. Front. Neurorobot. **7**, 21 (2013)
48. X.-X. Zhang, T. Deng, G.-Z. Jia, Nuclear spin-spin coupling constants prediction based on XGBoost and LightGBM algorithms. Mol. Phys. **118**(14), e1696478 (2020)
49. T.W. Cabral et al., Analysis of variance combined with optimised gradient boosting machines for enhanced load recognition in home energy management systems. Sensors **24** (2024). https://doi.org/10.3390/s24154965
50. T. Chen, C. Guestrin, Xgboost: a scalable tree boosting system, In *Proceedings of the 22nd ACM SIGKDD International Conference on Knowledge Discovery and Data Mining* (2016)
51. D. Zhang, Y. Gong, The comparison of LightGBM and XGBoost coupling factor analysis and prediagnosis of acute liver failure. IEEE Access **8**, 220990–221003 (2020)
52. K.L. Goh, A. Goto, Y. Lu, LGB-Stack: stacked generalization with LightGBM for highly accurate predictions of polymer bandgap. ACS Omega **7**(34), 29787–29793 (2022)
53. C. Jamotton, D. Hainaut, T. Hames, Insurance analytics with clustering techniques. Risks **12** (2024). https://doi.org/10.3390/risks12090141
54. S.V. Trepalin, A.V. Yarkov, Hierarchical clustering of large databases and classification of antibiotics at high noise levels. Algorithms **1**, 183–200 (2008). https://doi.org/10.3390/a1020183
55. B. Kalita et al., Learning to approximate density functionals. Acc. Chem. Res. **54**(4), 818–826 (2021)
56. S. Uhlmann, S. Kiranyaz, M. Gabbouj, Semi-supervised learning for Ill-posed polarimetric SAR classification. Remote Sens. **6**, 4801–4830 (2014). https://doi.org/10.3390/rs6064801
57. Y. Cho et al., SemiH: DFT Hamiltonian neural network training with semi-supervised learning. Mach. Learn.: Sci. Technol. **5**(3), 035060 (2024)
58. S.A. Meldgaard et al., Structure prediction of surface reconstructions by deep reinforcement learning. J. Phys.: Condens. Matter **32**(40), 404005 (2020)
59. P. Govindarajan et al., Learning conditional policies for crystal design using offline reinforcement learning. Digit. Discov. **3**(4), 769–785 (2024)
60. T. Hachaj, M. Piekarczyk, On explainability of reinforcement learning-based machine learning agents trained with proximal policy optimization that utilises visual sensor data. Appl. Sci. **15** (2025). https://doi.org/10.3390/app15020538
61. A. Ullah, Y. Chen, P.O. Dral, Molecular quantum chemical data sets and databases for machine learning potentials. Mach. Learn.: Sci. Technol. **5**(4), 041001 (2024)
62. T. Emmanuel et al., A survey on missing data in machine learning. J. Big Data **8**, 1–37 (2021)
63. *Compare the effect of different scalers on data with outliers*, 02 Mar 2025, https://scikit-learn.org/dev/auto_examples/preprocessing/plot_all_scaling.html.
64. R. Bakır, C. Orak, A. Yüksel, Optimising hydrogen evolution prediction: a unified approach using random forests, lightGBM, and bagging regressor ensemble model. Int. J. Hydrogen Energy **67**, 101–110 (2024)
65. J. Gibson, A. Hire, R.G. Hennig, Data-augmentation for graph neural network learning of the relaxed energies of unrelaxed structures. NPJ Comput. Mater. **8**(1), 211 (2022)
66. A. Chugh, *MAE, MSE, RMSE, Coefficient of Determination, Adjusted R Squared—Which Metric is Better?* 8 Dec 2020; 02 Mar 2025, https://medium.com/analytics-vidhya/mae-mse-rmse-coefficient-of-determination-adjusted-r-squared-which-metric-is-better-cd0326a5697e

67. XanaduAI, *GradDFT* (2023)
68. P. Kovács, P. Blaha, G.K.H. Madsen, Origin of the success of mGGAs for bandgaps. J. Chem. Phys. **159**(24) (2023)
69. D. Khan, M.L. Ach, O.A. von Lilienfeld, *Adaptive atomic basis sets*. arXiv, 2404.16942 (2024)
70. K.R. Stoltz, M.F. Borunda, Benchmarking DFT and supervised machine learning: an organic semiconducting polymer investigation. J. Phys. Chem. A **128**(4), 709–715 (2024)
71. Y. Zhang et al., Artificial neural networks assisted by DFT for boosting ligand design in ethylene tri-/tetramerization using silicon-bridged diphosphine (SBDP) ligands. Molecular Catalysis **576**, 114952 (2025)
72. N. Artrith, A.M. Kolpak, Understanding the composition and activity of electrocatalytic nanoalloys in aqueous solvents: a combination of DFT and accurate neural network potentials. Nano Lett. **14**(5), 2670–2676 (2014)
73. B. Kang, C. Seok, J. Lee, Prediction of molecular electronic transitions using random forests. J. Chem. Inf. Model. **60**(12), 5984–5994 (2020)
74. X. Tian et al., Screening for shape memory alloys with narrow thermal hysteresis using combined XGBoost and DFT calculation. Comput. Mater. Sci. **211**, 111519 (2022)
75. P. Xu et al., Machine learning aided design of polymer with targeted band gap based on DFT computation. J. Phys. Chem. B **125**(2), 601–611 (2021)
76. U. Saha, K. Debnath, S. Satapathi, *Screening of potential double perovskite materials for photovoltaic applications using agglomerative hierarchical clustering*. arXiv:2111.07557 (2021)
77. A. Chandrasekaran et al., Solving the electronic structure problem with machine learning. NPJ Comput. Mater. **5**(1), 22 (2019)
78. T. Zhang et al., Machine learning prediction of photocatalytic lignin cleavage of C-C bonds based on density functional theory. Mater. Today Sustain. **20**, 100256 (2022)
79. H. Mai et al., Rapid design of efficient Mn3O4-based photocatalysts by machine learning and density functional theory calculations. Adv. Energy Sustain. Res. 2400397 (2025)

The manufacturer's authorised representative in the EU is Springer Nature Customer Service Centre GmbH, Europaplatz 3, 69115 Heidelberg, Germany. If you have any concerns regarding our products, please contact ProductSafety@springernature.com

Printed and bound by CPI Group (UK) Ltd, Croydon, CR0 4YY
21/09/2026
02219966-0003